Enhanced Virtual Prototyping for Heterogeneous Systems

Muhammad Hassan • Daniel Große •
Rolf Drechsler

Enhanced Virtual Prototyping for Heterogeneous Systems

 Springer

Muhammad Hassan
DFKI GmbH
Bremen, Germany

Daniel Große
Johannes Kepler University of Linz
Linz, Austria

Rolf Drechsler
University of Bremen and DFKI GmbH
Bremen, Germany

ISBN 978-3-031-05576-8   ISBN 978-3-031-05574-4   (eBook)
https://doi.org/10.1007/978-3-031-05574-4

This Springer imprint is published by the registered company Springer Nature Switzerland AG
The registered company address is: Gewerbestrasse 11, 6330 Cham, Switzerland

To Khushboo and Maryam,
Marie,
and
Anna.

Preface

The successful co-design and verification of secure multi-disciplinary heterogeneous *systems-on-chips* (SOCs) with tight interactions between *hardware/software* (HW/SW) systems and their analog physical environment is an increasingly daunting task. In this regard, the emergence of *virtual prototypes* (VPs) at the abstraction of *electronic system level* (ESL) has modernized the design and verification of heterogeneous SOCs. A VP is essentially an executable abstract model of the entire HW platform and pre-dominantly created in C++-based system modeling language SystemC together with *transaction level modeling* (TLM) techniques and its mixed-signal extension SystemC AMS. The much earlier availability as well as the significantly faster simulation speed in comparison to the *register transfer level* (RTL) models and SPICE-level models are among the main benefits of VPs. Thus, virtual prototyping enables HW/SW co-design and verification very early in the design flow. Serving as reference for (early) embedded SW development and HW verification, the functional correctness and security validation of VPs are very important. Hence, a VP is subjected to rigorous functional verification. However, the modern VP-based verification flow still has weaknesses, in particular due to lack of methodologies to capture the complex interactions between digital and analog designs as well as unavailability of security validation techniques. This book proposes several novel approaches that cover varying verification aspects to strongly enhance the modern VP-based verification flow. The chapters of the book are essentially divided into four parts: The first part introduces a new verification perspective for VPs by using *metamorphic testing* (MT) as no reference models/value are needed for verification, unlike in modern VP-based verification flow. The second part enhances the code coverage closure methodologies in modern VP-based verification flow by considering *mutation analysis* and stronger coverage metrics like *data flow* coverage. The third part covers a set of novel, systematic, and lightweight functional coverage-driven verification methodologies to improve the coverage closure. The fourth and final part of the book showcases novel approaches to enable early security validation of VPs. All approaches are presented in detail and are extensively evaluated with several experiments that clearly demonstrate their effectiveness in strongly enhancing the modern VP-based verification flow.

Bremen, Germany   Muhammad Hassan
Linz, Austria   Daniel Große
Bremen, Germany   Rolf Drechsler
November 2021

Acknowledgments

First, we would like to thank the members of the research group for *computer architecture* (AGRA) at the University of Bremen as well as the members of the research department for *cyber-physical systems* (CPS) at the *German Research Centre for Artificial Intelligence* (DFKI) in Bremen. We appreciate the great atmosphere and stimulating environment. Furthermore, we would like to thank all co-authors of the papers which formed the starting point for this book: Hoang M. Le, Vladimir Herdt, Mingsong Chen, and Mehran Goli. We especially thank Karsten Einwich and Thilo Vörtler from COSEDA Technologies GmbH for many interesting discussions and successful collaborations.

Contents

List of Algorithms

List of Figures

List of Tables

Chapter 1
Introduction

Internet-of-Things (IOT) and *5G* (fifth generation technology standard for broadband cellular networks) have enabled a plethora of possibilities which were once unimaginable. While *5G* provides the high-speed connectivity and ubiquitous coverage, it is the smart IOT devices that gather and transport the data that fuel the promise and potential of IOT. These smart devices are a prime example of heterogeneous *System-On-Chips* (SOCs), which comprise two parts: (1) Mixed-Signal *Hardware* (HW) where analog world meets the digital world, (2) and *Software* (SW), the invisible layer that connects us to the physical reality. Heterogeneous SOCs are among the fastest growing market segments in the electronics and semiconductor industry. Driven by growth opportunities in various application domains, many semiconductor vendors are adapting and shifting their focus from separate *Integrated Circuits* (ICs) performing one functionality, toward a more integrated solution of *Radio Frequency* (RF) and high-performance *Analog/Mixed-Signal* (AMS) designs. Due to this industry shift, most SOCs today are heterogeneous containing analog sensors, mixed-signal converters, digital processors running SW on top, and RF transceivers, tightly integrated on a single die. While this shift has resulted in high-performance, efficient, and low-area devices, e.g., Apple M1 SOC [6], it has significantly increased the efforts required to develop and verify these highly complex devices and achieving the required *Time-To-Market* (TTM) simultaneously.

The first challenge in this regard is the HW and SW dependency. Conventionally, HW and SW were developed in isolation and only met each other at the late integration and testing stages. As a consequence, a sequential dependency between HW and SW development phases always existed as shown in Fig. 1.1a. Hence, SW could only be tested properly once the first silicon prototypes of the SOC were available. In particular, HW dependent SW such as device drivers and low-level kernel code could only be written after the silicon design had been completed. One solution to lessen the TTM widely adopted by industry is to move away from complete in-house HW development and instead use larger amounts of pre-

© The Author(s), under exclusive license to Springer Nature Switzerland AG 2023
M. Hassan et al., *Enhanced Virtual Prototyping for Heterogeneous Systems*,
https://doi.org/10.1007/978-3-031-05574-4_1

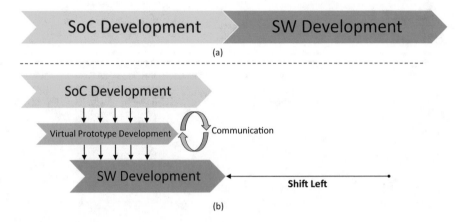

Fig. 1.1 Early SW development leveraging shift left concept

verified third-party *Intellectual Property* (IP). It allows them to focus more on the *Unique Selling Point* (USP) of their HW and SW. With the increased SW feature-rich functionality and complex interactions between different HW components, the design complexity has increased manyfold making the design verification of heterogeneous SOCs an increasingly daunting task.

The second challenge in verification of heterogeneous SOCs is the slow joint simulation speed of *Register Transfer Level* (RTL) and SPICE (Simulation Program with Integrated Circuit Emphasis) models for the digital and analog/RF part of the SOC [9]. Traditionally, analog/RF verification methodology was ad hoc by nature and these IPs were always verified by separate teams. It was driven by directed tests run over sweeps, corners, and Monte Carlo analysis. Unfortunately, it has not changed much until now and creates a bottleneck in the design and verification process. On the other hand, digital IPs had formalized verification methodologies used by separate teams. This included executable verification plans, constrained-random stimulus generation [119], testbench automation, assertions, and coverage metrics. These pre-verified analog, RF, and digital IPs were later on integrated together in a mostly digital SOC design and SW was executed on top to test if everything works as expected. However, due to multi-functional nature of the heterogeneous SOCs, the analog and RF IPs in particular have become very complex with significant digital control logic. Furthermore, the interaction of analog/RF and digital IPs has increased significantly, in particular when SW running on top is considered as well [77]. Hence, the traditional verification approaches are no longer adequate. The joint simulation, while slow, is still considered a golden standard because of high accuracy and cannot be ignored. However, it is too slow for chip-level simulations, unless it is used extremely selectively.

Third, design space and architectural exploration is restricted. Given the system requirements, finding the optimal system configuration is a tiring task. Each parameter change, no matter how minor, can cost the designer a couple of days in design phase. Hence, the exploration is very selective.

Lastly, the widely accepted practice of third-party IP integration has made the modern SOCs notoriously insecure as they can be used as a vehicle for malice. For example, the recent security compromise of SOCs using Intel's microprocessor IPs and Actel ProASIC3 IPs. The former were exploited by *"Meltdown and Spectre"* vulnerabilities [79, 89] and the latter with JTAG vulnerability [108]. This highlights the fact that legitimate commercial off-the-shelf IPs may manipulate or assist in manipulating secure user information in ways that their users neither expect nor appreciate. Moreover, SOC cannot be patched after fabrication, and hence, it requires a new design.

According to an estimate, the traditional functional verification of heterogeneous SOCs can take up to 70% of the project time. Yet, result in many re-spins, which mainly stem from mixed-signal verification issues [77]. This primarily happens because of weaker verification methodologies, tools, or libraries which are not able to detect *avoidable* HW and SW errors early in the design phase. Therefore, tools, libraries, and complex AMS verification methodologies are required which can (1) Remove the *HW/SW serialized dependency* in heterogeneous SOCs design and verification and instead enable HW/SW co-design (2) *Increase simulation speed* and provide good accuracy simultaneously (3) Enable *rapid design space and architectural exploration* (4) *Provide early security validation*

To summarize, the successful co-design of secure multi-disciplinary heteroge-neous SOCs exhibiting tight interactions between HW/SW systems and their analog physical environment is challenging. As TTM becomes shorter, the ability to model and simulate complex heterogeneous SOCs where digital HW/SW is functionally intertwined with AMS IPs, i.e., RF interfaces, power electronics, sensors, and actuators, becomes more and more essential. If such overall system and architectural level models are available as early as possible in the design cycle, the architecture exploration and design issues, as well as security leaks, will be dramatically reduced.

1.1   Electronic System-Level Design and Verification

In this regard, the emergence of *Virtual Prototypes* (VPs) at the abstraction of *Electronic System Level* (ESL) has modernized the design and verification of heterogeneous SOCs in many ways [10, 33, 43, 45, 63–66, 91, 94]. The high-level idea is to create an abstract reference model of the SOC from written specifications. As a consequence, an executable description is made available which is used as a golden reference for both HW and SW development. Hence, *virtual prototyping* provides SW developers and system architects with an environment for SW development, architecture exploration, or HW/SW co-design. In a complete SOC design flow, virtual prototyping falls between the functional level and the implementation level as shown in Fig. 1.2.

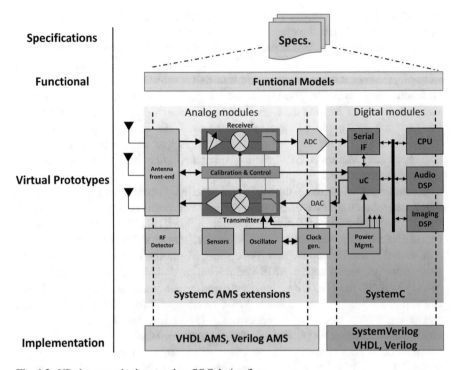

Fig. 1.2 VP placement in the complete SOC design flow

At this abstraction, virtual prototyping leverages the *Shifting Left* concept because of its early availability as shown in Fig. 1.1b, i.e., the HW architecture design and SW development flow is done in parallel and interleaved manner. Consequently, this HW/SW co-design and verification facilitates integration, reduces engineering costs, and shortens TTM. Additionally, the interleaving of HW/SW flow enables a feedback path where observations from SW space can be used to improve and optimize the HW. Furthermore, VPs are high-speed fully functional SW models of physical HW systems which can model a complete electronic system while offering a good trade-off between design accuracy and simulation speed [67, 68]. The high-speed enables *rapid design space and architecture exploration*. Additionally, this SW nature of the VPs allows cost-effective powerful debugging mechanisms via non-intrusive introspection into the entire heterogeneous SOC which are almost unthinkable on a real HW system [39, 42]. This leads to a higher quality of the product and a lower supporting effort. For the ESL design, the state-of-the-art modeling language is C++-based system modeling language SystemC together with *Transaction-Level Modeling* (TLM) techniques (IEEE Standard 1666) [73] and its mixed-signal extension SystemC AMS [11] with pre-dominantly *Timed Data Flow* (TDF) model development [9, 10, 12, 44, 94]. A detailed description on these modeling languages can be found in Chap. 2.

To summarize, the adoption of SystemC/AMS-based VPs has led to significant improvements on the design and verification of heterogeneous SOCs. The much earlier availability as well as the significantly faster simulation speed in comparison to the RTL models (for digital) and SPICE-level models (for AMS) is among the main benefits of VPs. Thus, virtual prototyping enables HW/SW co-design and verification very early in the design flow. Serving as reference for (early) embedded SW development and HW verification, the functional correctness and security validation of VPs is very important. Hence, a whole VP as well as its individual components, i.e., high-speed RF, AMS, and digital IPs, is subjected to rigorous functional verification.

The functional verification of the VPs can be broadly determined using methodologies from two categories: (1) formal verification methods and (2) simulation-based verification methods. Both of them are proven effective in the digital design verification, and however, former is still in infancy for AMS SOC verification [8, 49, 109, 110, 120–123]. Formal verification checks the correctness of the *Design Under Verification* (DUV) based on certain formal methods of mathematics, e.g., model checking, equivalence checking, etc. This kind of verification methods implies one prerequisite on the system to be verified: the number of states, value candidates, or the time points must be finite. Unlike the digital signals, the analog signals are continuous and potentially have infinite values. Their discretization may generate a huge number of states and points of time which may lead to state space explosion problem [95]. As a consequence, practical applications of the formal verification method for the large-scale AMS SOC are still in its infancy.[1]

On the other hand, despite the recent progress in formal verification of SystemC/AMS models (see, e.g., [22, 24, 45, 49, 69, 70, 85, 86, 120]), simulation-based verification is still the method of choice for SystemC/AMS-based VPs, thanks to its scalability and ease of use. A general simulation-based AMS VP verification environment is shown in Fig. 1.3. It follows the principles of *Directed Testing*. Basically, the *textual specifications* are used to manually create a set of stimuli (*VP Testsuite*), which is applied to the AMS DUV (which can be either a whole VP, a set of components, or a single component) to test specific scenarios. For each stimulus, the actual behavior is checked against the expected behavior (e.g., specified by

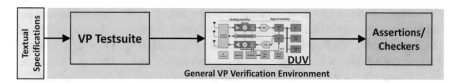

Fig. 1.3 General AMS VP verification environment

---

[1] Formal methods will not be discussed in detail as they are not the focus of this book.

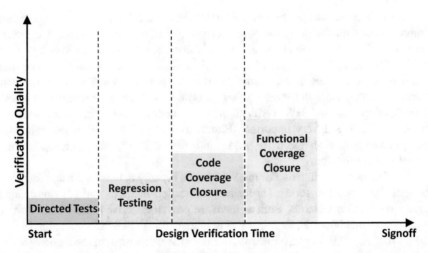

Fig. 1.4 Four stages of modern simulation-based VP verification flow

reference outputs in the form of *assertions/checkers* or temporal properties). If the assertions/checkers fail, the DUV is declared faulty.

While this general simulation-based verification flow (directed testing) is important for initial verification of a simpler DUV, it is not effective for complex designs and thorough verification [23]. Therefore, modern VP verification flow is nowadays widely used with various methodologies to complement the general verification flow. This is shown in Fig. 1.4. *x-axis* shows four colored bars to represent the methodologies of modern verification flow from *start* to *signoff*, and *y-axis* represents the corresponding verification quality achieved by each set of methodologies. The high-level idea is to start with *directed testing* as described earlier. This is shown as gray color bar. The verification quality is very poor during this stage as the test stimuli only verifies specific scenarios. Afterwards, the initial set of stimuli from directed testing is used in addition to constrained-random techniques [50, 119] for *regression testing*. Regression testing effectively captures (1) bugs introduced during DUV development and (2) code coverage (exercised lines of code) baselines and trends. This is represented by orange color bar. The verification quality achieved by regression testing is better than directed testing but still poor as the focus is to develop stimuli. Afterwards, leveraging the code coverage baselines, verification flow transitions to achieve code coverage closure, and hence, the verification quality increases. This is represented by blue color bar. At the end, the coverage metric is changed to functional coverage (exercised features of a DUV) and the stimuli is improved to achieve closure (yellow color bar). At this point, the verification quality is considered really good and verification signoff is done. However, the modern VP verification flow still has weaknesses which lead to poor quality test stimuli and VP. The weaknesses are briefly discussed below:

1. During regression testing, the availability of reference models for output comparison is always a major challenge. For complex DUVs, significant effort is needed to specify the reference behavior in an executable way. In particular, the interaction of analog designs and digital logic has increased significantly in modern AMS SOCs. Formalizing such interactions is non-trivial and very time-consuming. Therefore, verification approaches are required which do not require reference models to determine the correctness of DUVs.

2. While the existing code coverage closure methodologies are good, they fall short in two aspects. First, they use only weak coverage metrics, e.g., statement coverage and branch coverage, etc. The weak coverage metrics are insensitive to variable interactions in the VP, i.e., how the computations done in one part of VP affect the other parts. Second, they do not consider the HW/SW interactions that result in a large number of IP integration issues later on. As a consequence, comprehensive verification of VPs cannot be achieved. Hence, a more holistic approach is required which leverages system-level software in combination with stronger code coverage metrics to verify the VPs.

3. For thorough verification of the DUV, tracking of verification progress is required. In digital design verification, functional coverage, in particular, is the metric used for this task since it allows to measure if all specifications of the design have been verified [96]. While functional coverage is very well understood in digital design (see, e.g., [46, 93, 114]), this is not the case for AMS [37, 54] as continuous signals and their change over time are much harder to capture. Nevertheless, methodologies driven by functional coverage have been also considered for AMS. However, the existing methodologies for AMS suffer from three major shortcomings: (1) they are not systematic, (2) they only consider linear behaviors of AMS systems, and (3) complex mathematical models are required to capture specifications. This is clearly a problem as the true complexity and corner-cases stem from non-linear and unstable (overshooting and undershooting) behaviors. Moreover, capturing these behaviors is non-trivial and requires a lot of effort. Therefore, advanced verification methodologies are required to systematically verify the different behaviors of DUVs.

4. Security is one of the most burning issues in embedded system design nowadays. The majority of strategies to secure embedded systems are being implemented in software. However, a potential hardware backdoor that allows unprivileged software access to confidential data will render even the perfectly secure software useless. As the underlying SOC cannot be patched after deployment, it is very critical to detect and correct SOC hardware security issues in the design phase. To prevent costly fixes in later stages, security validation should start as early as possible. However, modern VP-based verification flow lacks early security validation methodologies.

In the next section, the contributions of this book are discussed which strongly enhance the quality of modern VP verification flow.

1.2　Book Contribution

The VP-based modeling and verification is heavily used today [43, 57–59, 63–66, 91, 94, 119]. It has significantly improved the overall SOC design and verification process. However, as mentioned earlier, the modern VP-based verification flow still has weaknesses. This book proposes several novel approaches and methodologies to strongly enhance the modern VP verification flow. The contributions are broadly proposed in four major areas after the *directed tests* as shown in Fig. 1.5. First, this book proposes a new verification perspective for VP verification: *Metamorphic Testing* to effectively verify the VPs without the need for reference models. It is represented by green color bar in Fig. 1.5. Then, this book proposes strong enhancements up to 30% in code coverage and functional coverage closure methodologies (blue and yellow color bars, respectively). Lastly, security validation of the VPs is introduced (pink color bar) to identify security issues early on in the design phase. In this regard, we focus on digital as they have compromised in the recent years, e.g., security compromise of SOCs using Intel's microprocessor IPs (*"Meltdown and Spectre"* vulnerabilities [79, 89]) and Actel ProASIC3 IPs (JTAG vulnerability [108]). Hence, identifying such security flaws early on can be critical for the SOC. A more detailed overview of the book contributions is shown on the left side of Fig. 1.6. The four areas of contribution use general VP verification environment as base and build strongly enhanced verification environments on top:

1. AMS Metamorphic Testing Environment
2. AMS Enhanced Code Coverage Verification Environment
3. AMS Enhanced Functional Coverage Verification Environment
4. Digital Early Security Validation

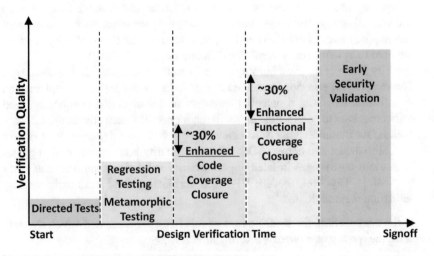

Fig. 1.5 Proposed enhancements in VP verification flow

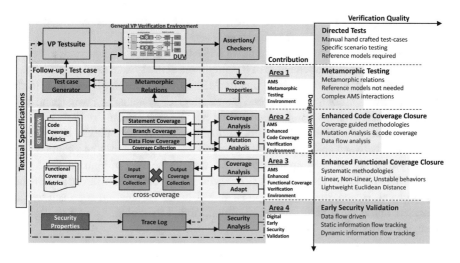

Fig. 1.6 Contributions of book: enhanced VP verification flow

Each contribution area significantly increases the verification quality of VP. This is reflected on *x-axis* in Fig. 1.6 (on right side). *y-axis* represents corresponding verification environment and methodologies from *start* to *signoff* in VP verification flow. The contributions achieve a high quality testbench and a thoroughly verified VP and will be detailed in the following:

1.2.1 Contribution Area 1: AMS Metamorphic Testing Environment

The first contribution of this book overcomes the *first weakness—availability of reference models*. The contribution introduces a new verification perspective for VPs by using *Metamorphic Testing* (MT). MT verifies a DUV by considering the already available test-cases from directed or regression testing (termed *base test stimuli*) to generate new test-cases (termed *follow-up test stimuli*) without the need of a reference model. MT does that just by using and relating available test-cases with newly created test-cases. The central element of MT is a set of *Metamorphic Relations* (MRs), which describes the relation of the inputs and outputs of consecutive DUV executions using core properties of DUV. Since reference models are not required in MT, it effectively complements *regression testing*. As a consequence, the identified MRs ensure the DUV correctness across multiple DUV versions during design and verification. The proposed AMS Metamorphic Testing Environment is shown in Fig. 1.6 in green background. It comprises (1) test case generator, (2) metamorphic relations, and (3) core properties of DUV. The high-level idea is to create MRs using the core properties of DUV. Afterwards, use the MRs in the test-case generator module to create new test-cases which can reveal bugs.

Hence, in this book, we present a novel MT-based verification approach to verify the linear and non-linear behaviors of RF amplifiers at the system level. For the class of RF amplifiers, we identify high quality MRs to verify the linear and non-linear behaviors. Furthermore, we go beyond pure analog/RF systems, i.e., we expand MT to verify AMS systems. High-speed RF or pure analog systems only experience one type of signals. This is clearly not sufficient as the true complexity stems from interactions between analog and digital components. Consequently, as a challenging AMS system, we consider an industrial PLL and devise a set of high quality MRs. These MRs allow to verify the PLL behavior at the component level and at the system level. Hence, this book proposes an MT-based verification approach which considers analog-to-analog, analog-to-digital, digital-to-analog, as well as digital-to-digital behavior.

1.2.2 Contribution Area 2: AMS Enhanced Code Coverage Verification Environment

As a second contribution, this book strongly enhances the code coverage closure methodologies in modern VP-based verification flow. The methodologies achieve a significant enhancement of up to 30% in verification quality. The proposed AMS Enhanced Code Coverage Verification Environment is shown in Fig. 1.6 in blue background. It comprises various code coverage metrics and novel coverage analysis. Additionally, it leverages mutation library and *Mutation Analysis* to achieve a high quality VP.

As mentioned earlier, the much earlier availability and the significantly faster simulation speed of VP-based designs in comparison to RTL are among the main benefits of SystemC/AMS-based VPs. These enable HW/SW co-design, and hence, enable *Software Driven Verification* (SDV), which has the promise to significantly reduce the overall time and effort for the task of IP integration and verification. With the help of SystemC VPs, SW tests to verify the (new) integrated IP blocks and the HW/SW integration can be developed in an early design stage and reused in the subsequent steps. For this purpose, we propose in this book a novel *quality-driven methodology* based on mutation analysis. By elevating the main concepts of mutation-based qualification to the context of SDV, our methodology is capable to detect serious quality issues in the SW tests. At its heart is a novel consistency analysis, which measures the coverage of the IP in HW/SW co-simulation in a lightweight fashion and relates this coverage to the SW test results to provide clear feedback on how to further improve the quality of tests. However, although a necessary step, statement and branch coverage in the context of SDV have some well-known limitations in their capability to detect bugs as well as to reflect the thoroughness of verification. They fall short when considering the interactions between different elements (variables) in a VP.

In this regard, *Data Flow Testing* (DFT) enhances the verification quality by considering how one syntactic element can affect the computation of another. Hence, in this book, we propose to apply DFT for SystemC/AMS VPs as the modern VPs are not digital-only anymore, rather they are multi-functional AMS SOCs. We first develop a set of SystemC/AMS specific coverage criteria for DFT. This requires to consider (1) SystemC semantics of using non-preemptive thread scheduling with shared memory communication and event-based synchronization and (2) SystemC AMS semantics of signal flow in general and timed data flow models in particular. Afterwards, we explain how to automatically compute the data flow coverage results for a given VP using a combination of static and dynamic analysis techniques. The coverage results provide clear suggestions for the verification engineer to add new test-cases in order to improve the coverage result.

1.2.3   Contribution Area 3: AMS Enhanced Functional Coverage Verification Environment

The third contribution of this book strongly enhances the functional coverage closure methodologies in modern VP verification flow by increasing the verification quality up to 30%. The AMS Enhanced Functional Coverage Verification Environment is shown in Fig. 1.6 in yellow background. It comprises coverage collection bins on input and output of DUV as well as coverage analysis. Furthermore, it introduces *adapt* module to guide the stimuli generation automatically.

This book proposes a functional coverage-driven verification approach as a systematic solution for the class of RF amplifiers to verify the linear and non-linear behaviors. It elevates the main concepts of digital functional coverage to the context of SystemC AMS in particular and system-level simulations in general. To enable AMS functional coverage-driven verification, it introduces two coverage refinement parameters on input and output side, to systematically generate input stimuli and capture specifications. At the heart of the approach is the coverage analysis which measures the functional coverage of the DUV and provides clear feedback to the verification engineer to reach coverage closure. However, the coverage refinement parameters need to be adjusted manually which creates a bottleneck for complex systems and unstable behaviors.

In this regard, a *Lightweight Coverage-Directed Stimuli Generation* (LCDG) approach is considered. CDG is a verification methodology that aims to reach coverage closure automatically by using coverage data and mathematical functions to direct the next iterations of test stimuli generation. At the heart of the proposed LCDG approach is a coverage analysis which leverages functional coverage data of input, output, and checkers, in combination with *Euclidean Distance* to achieve coverage closure. The much simpler *Euclidean Distance* in contrast to *Bayesian*

Networks or complex probability distribution functions makes it lightweight. In case of coverage holes, the analysis automatically adjusts the coverage refinement parameters to increase the coverage of the DUV. As a consequence, these lightweight and systematic approaches ensure efficient convergence and thorough verification of the VP.

1.2.4   Contribution Area 4: Digital Early Security Validation

The last contribution of this book is the early security validation of the functionally verified VP. The Digital Early Security Validation Environment is shown in Fig. 1.6 in pink background. It comprises three major components: (1) security properties, (2) trace logs, and (3) combination of static and dynamic security analysis. Leveraging these components, this book proposes a novel approach to SOC security validation at the system level using VPs. At the heart of the approach is a scalable static information flow analysis that can detect potential security breaches such as data leakage and untrusted access, *confidentiality* and *integrity* issues, respectively.

Furthermore, this book complements the static approach by presenting the dynamic information flow analysis for VPs. In particular, it looks into the IP isolation security feature, which is widely used nowadays, e.g., *secured Memory Mapped IOs* (MMIOs), or secured address ranges in case of memories, are marked as non-accessible. One way to provide strong assurance of security is to define isolation as information flow policy in hardware using the notion of non-interference. The proposed approach allows to validate the run-time behavior of a given SOC implemented using VPs against security threat models, such as information leakage (*confidentiality*) and unauthorized access to data in a memory (*integrity*).

1.2.5   Contribution Summary

All the aforementioned contributions have been implemented and extensively evaluated with several experiments on real-world industrial models. Detailed description and results will be presented in the following chapters. In summary, these contributions strongly enhance the modern VP-based verification flow as demonstrated by the experiments. One of the main benefits is the drastically improved verification quality in combination with a significantly lower verification effort. On the one hand, this reduces the number of undetected bugs and increases the overall quality of AMS SOC. On the other hand, a high quality VP testsuite is created which can be used for verifying the lower abstractions.

1.3 Book Organization

The book is structured as follows: Chap. 2 provides a relevant background information of SystemC TLM and SystemC AMS. Afterwards, Chap. 3 presents metamorphic testing approaches for the verification of AMS VPs. Then, Chap. 4 presents the enhanced code coverage closure methodologies using stronger coverage metrics for achieving high quality VP. Chapter 5 presents enhanced functional coverage closure methodologies to capture linear, non-linear, and unstable behaviors of RF amplifiers. Chapter 6 considers security aspects of the VP statically and dynamically to enable early security validation. Finally, Chap. 7 concludes the book.

Chapter 2
Preliminaries

This chapter provides a background on the common topics to keep the book self-contained and avoid duplication. As the focus of this book is VP verification using novel methodologies, it makes sense to briefly discuss *SystemC* and its mixed-signal extension which will be used throughout the book. As motivated in the introduction, *SystemC* is the state-of-the-art modeling language for VPs. *SystemC* offers a single, unified design and verification language that spans HW and SW to express architectural and system-level attributes of AMS SOCs in the form of standard C++ classes. The architecture of *SystemC* is shown in Fig. 2.1. It comprises SystemC core language together with its *Transaction-Level Modeling* (TLM) techniques (IEEE Standard 1666−2011) [73] and its mixed-signal extension SystemC AMS (IEEE Standard 1666.1−2016) [11].

First, SystemC core language and its TLM standard are briefly discussed in Sect. 2.1. Then, SystemC AMS and its *Models Of Computation* (MOC) are introduced in Sect. 2.2. Please note that parts of the following preliminaries on *SystemC* also already appeared in similar form in the papers mentioned in the introduction which are co-authored by the book author.

2.1 SystemC

SystemC is a library built on top of standard C++ language. It enables modeling of electronic systems comprising of both HW and SW, via its event-driven simulation kernel. To achieve this, SystemC allows to implement C++ classes (module) and functions (processes) which are supervised via a non-preemptive process scheduler. The scheduler is responsible for process scheduling w.r.t. simulated time, synchronization of processes, and communication between processes via events. These processes define the behavior of the SystemC model and events are notified at specific points in simulated time. They are encapsulated in a module that handles

© The Author(s), under exclusive license to Springer Nature Switzerland AG 2023
M. Hassan et al., *Enhanced Virtual Prototyping for Heterogeneous Systems*,
https://doi.org/10.1007/978-3-031-05574-4_2

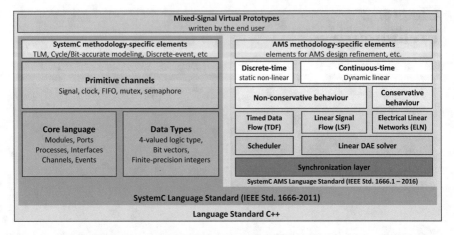

Fig. 2.1 SystemC architecture with TLM and AMS extensions [1, 71]

the structural and connectivity aspects of the electronic systems via ports, channels, interfaces, etc. The architecture of SystemC is shown in Fig. 2.1 (blue background). Broadly, it contains four components: (1) core language, (2) data types, (3) primitive channels, and (4) methodology-specific elements.

In the next section, basics of SystemC are briefly discussed with the help of an example. Afterwards, TLM-based communication is presented, and at the end the simulation semantics are discussed.

2.1.1 Basics

To understand the basics of modeling HW using SystemC, a *counter* example is shown in Listing 2.1. The *counter* counts up on every positive edge of the clock signal and prints only even values. The *counter* has the following elements:

1. **Module** is the basic building block of the SystemC model, defined as *SC_MODULE (counter)* (Line 1). A bigger system, e.g., SOC, can be represented by instantiating multiple modules inside a module. A module contains ports, interfaces, channels, processes, events, etc.
2. **Ports, channels, and interfaces** are used for communication between modules or between processes. In this case, three input ports, *sc_in_clk clock*, *sc_in<bool> rst*, and *sc_in<bool> en* and one output port *sc_out<sc_uint<4> > count_out* are defined (Lines 2–5).
3. **Processes** implement the behavioral functionality of the system. There are three types of processes, *SC_THREAD*, *SC_METHOD*, and *SC_CTHREAD*. In this example, there are two processes: *count_up* (Line 20) and *print_count* (Line 36).

The processes are sensitive to events which trigger their execution, e.g., *count_up* is sensitive to the positive edge of the *clock* (Line 13).

4. **Events** enable concurrency and event-based synchronization for HW models. In this case, process *print_count* waits for event *print_event*, *wait(print_event)* (Line 38). As soon as the event is notified from process *count_up*, the process prints the counter value.

5. **Clocks** are special signals that run periodically and can trigger clocked processes, e.g., process *count_up* is sensitive to positive edge of *clock* signal.

6. **Hardware modeling** is done using various semantics, e.g., *sc_signal*, *sc_logic*, *sc_bv*, etc.

7. **Software modeling** is done using the communication primitives, *sc_mutex* and *sc_semaphore*.

For more details regarding the SystemC basics and usage, please refer to *SystemC User Guide* [71]. In the next section, *Transaction-Level Modeling* (TLM) is briefly discussed.

```
1   SC_MODULE (counter) {
2     sc_in_clk clock ; // Clock input of the design
3     sc_in<bool> rst ; // active high, synchronous Reset input
4     sc_in<bool> en;   // Active high enable signal for counter
5     sc_out<sc_uint<4> > count_out; // 4 bit vector output of the
                counter
6
7     sc_uint<4> count;
8     sc_event print_event; // Event to print only even numbers
9
10    SC_CTOR(counter) {
11      // Clock positive edge sensitive
12      SC_THREAD(count_up);
13      sensitive << clock.pos();
14
15      SC_THREAD(print_count);
16    } // End of CTOR
17
18
19    // counter logic
20    void count_up () {
21      while (true) {
22        wait();
23        if (rst.read()) {
24          count = 0;
25          count_out.write(count);
26        } else if (en.read()) {
27          count++;
28          count_out.write(count);
29          if ((count % 2) == 0)
30            print_event.notify();
31        }
32      }
33    } // Function end
34
35    // prints value\texttt{}
36    void print_count () {
37      while (true) {
38        wait(print_event);
39        cout<<"Counter Value: "<< count_out.read()<<endl;
40      }
41    }
42  }; // End of Module
```

Listing 2.1 SystemC example

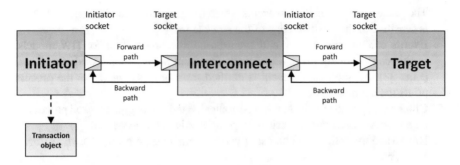

Fig. 2.2 TLM: initiator, target, socket, and interconnect

2.1.2 Transaction-Level Modeling (TLM)

Transaction-Level Modeling (TLM) in general, and TLM-2.0,[1] provides an abstracted approach to model communication between SystemC modules at high-speed (in comparison to pin-interfaces). This way, TLM provides a platform for SW development where the designer does not have to worry about the low-level semantics of ports. It is least concerned with the implementation of modules and threads within a module, rather it is concerned with how different modules communicate. TLM provides several interfaces, sockets, generic payload, and utilities, for communication (depending on the application).

The interfaces can be blocking and non-blocking, *Direct Memory Interface* (DMI), or the debug transport interface. The blocking and non-blocking interfaces provide different levels of timing details. The blocking interface only models the start and end of a transaction using single function call and does not expect a return value, whereas the non-blocking interface requires multiple function calls for a single transaction and has a return value. DMI provides a direct access to a memory area, hence, bypassing interface function calls. The debug transport interface provides introspection of the memory without passing the simulation time. DMI is intended to speed up simulation during normal transactions, whereas the debug transport interface is exclusively intended for debug.

TLM uses sockets to pass transactions between *initiator* and *target* (see Fig. 2.2). An *initiator* (e.g., CPU) creates and sends out the transactions via initiator socket, whereas the *target* (e.g., Memory) receives the incoming transactions through target socket and acts as a final destination of the transaction. Multiple modules are connected via an *interconnect* (bus) which acts as a router for the transactions. The generic payload is an important part of TLM as it enables interoperability between models. It holds its significance particularly for memory-mapped buses, and however, it has the capability to be used as the basis for modeling other protocols. Its attributes include *command*, *address*, *data*, *response status*, etc. The

---

[1] TLM will refer to TLM-2.0 from this point onward.

forward path in Fig. 2.2 refers to the passing of transaction object from initiator to target via interconnect, whereas the backward path is in the opposite direction. Finally, the TLM can be implemented in two coding styles: *Loosely Timed* (LT) and *Approximately Timed* (AT). LT style focuses on the functionality at high-speed, and hence, it keeps the track of time using simple techniques. On the other hand, AT is useful for architectural exploration and performance analysis as it includes approximate timing. AT models are significantly slower than LT model, and however, they are still significantly faster than RTL.

Some elements will be discussed in more detail in the following chapters. For in-depth reading, please refer to [71].

2.2   SystemC AMS

SystemC together with TLM is suitable for purely digital models. However, it does not offer efficient ways to design *Analog/Mixed-Signal* (AMS) and signal processing systems at the functional and architectural level. This drawback is fixed by the AMS extension of SystemC, called *SystemC AMS* (shown in Fig. 2.1 on right side with yellow background). They are built on top of SystemC standard. It allows high-speed AMS simulations up to $1000000\times$ in comparison to SPICE simulations (Fig. 2.3) and can be used for architectural exploration, integration verification, and virtual prototyping. Additionally, SystemC AMS provides new small-signal frequency-domain analysis techniques, data types, and AMS signals tracing.

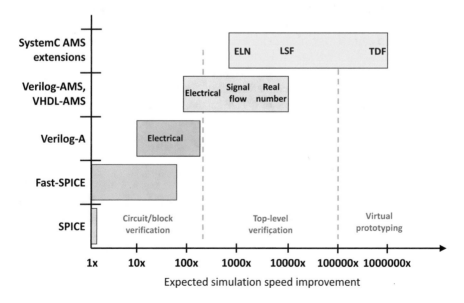

Fig. 2.3 Simulation speed of modeling languages in comparison to SPICE [9]

In the next section, we briefly discuss the supported *Models Of Computation* (MOC) by SystemC AMS, followed by *Timed Data Flow* MOC, which is the recommended MOC for AMS VP.

2.2.1   Models of Computation (MOC)

SystemC AMS extension offers three different MOC to support AMS VPs: (1) *Timed Data Flow* (TDF), (2) *Linear Signal Flow* (LSF), and (3) *Electrical Linear Networks* (ELNs). The three MOCs are shown in Fig. 2.1. TDF is useful for modeling and simulation of models at functional and architecture level in discrete time. LSF focuses on modeling of control systems in continuous time, and ELN allows modeling of electrical loads and buses at high frequencies in continuous time.

2.2.2   Timed Data Flow (TDF)

The TDF MOC is the recommended MOC for creating AMS models in virtual prototyping. TDF defines time domain processing and is used to model the pure algorithmic or procedural description of the underlying design. Because of earlier availability and significantly faster simulation speed (see Fig. 2.3), the TDF models provide a design refinement methodology and enable early verification for AMS systems. Multiple TDF models connect together to form a TDF cluster. The TDF models in a cluster communicate with each other using TDF channels and TDF ports. The behavior of a TDF model is described in predefined functions: (1) *processing*, (2) *initialize*, (3) and *set_attributes*. *Processing* implements the functional behavior, *initialize* specifies the initial conditions of the model, if any, and *set_attributes* describes the timing information of the TDF ports, e.g., timesteps. In the next section, a small example of a TDF model,: second-order passive *Low Pass Filter* (LPF), is presented [58].

Example: Low Pass Filter (LPF)

A structural model of second-order LPF is shown in Fig. 2.4, and it consists of a resistor (R), a capacitor (C), and an inductor (L). The corresponding values are $R = 4.3$ KΩ, $L = 470$ mH, and $C = 0.047$ µF. LPF is designed to allow signals with

Fig. 2.4 Second-order LPF

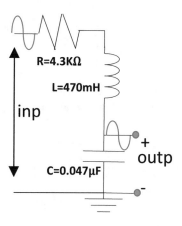

frequency lower than a certain cutoff frequency (F_c) and attenuate the signals with frequency higher than cutoff frequency. The cutoff frequency $F_c = 1$ KHz. The filter is implemented in SystemC AMS TDF MOC, where its Laplace transfer function (described by the numerator and denominator coefficients) is shown in Eq. 2.1.

$$H(s) = \frac{1}{LCs^2 + RCs + 1} \tag{2.1}$$

The TDF model of LPF is shown in Listing 2.2. SystemC AMS provides dedicated solver for continuous time linear transfer functions in Laplace domain under the class *sca_tdf::sca_ltf_nd*, defined here as *ltf* (Listing 2.2 Line 22). SCA_CTOR macro defines the constructor (Line 8). The member function *set_attributes* defines the time step of TDF module activation (Line 10). The library function *initialize* sets the initial values of the member variables, i.e., *num* and *den* (Line 14). This is also reflected in Eq. 2.2 with component values replaced.

$$H(s) = \frac{1}{22e^{-9}s^2 + 202.1e^{-6}s + 1} \tag{2.2}$$

The callback method *processing* defines the behavior of LPF (Line 21). The *num*, *den*, and *inp* are given to the solver (*ltf*) as inputs, and *ltf* returns the continuous output and assigns it to *outp*. For more details and examples of SystemC TDF models, please refer to [1, 13].

```
1  |  SCA_TDF_MODULE (lpf_tdf) {
2  |    sca_tdf::sca_in <double> inp;     // TDF input port
3  |    sca_tdf::sca_out <double> outp;    // TDF output port
4  |
5  |    sca_tdf::sca_ltf_nd ltf;           // laplace transform function
6  |    sca_util::sca_vector <double> num, den;  // coefficients
7  |
8  |    SCA_CTOR (lpf_tdf) { } // constructor
9  |
10 |    void set_attributes () { // executed only once in start
11 |      set_timestep(1.0, SC_UC); // time between activations
12 |    }
13 |
14 |    void initialize () {  // executed only once in start
15 |      num(0) = 1.0;
16 |      den(0) = 1.0;
17 |      den(1) = 202.1e^{-6};
18 |      den(2) = 22e^{-9};
19 |    }
20 |
21 |    void processing () { // activated at each activation
22 |      outp.write ( ltf(num, den, inp) );
23 |    }
24 |
25 |  };
```

Listing 2.2 SystemC AMS second-order LPF implementation [58]

Chapter 3
AMS Metamorphic Testing Environment

This chapter introduces a new verification perspective for AMS VPs that does not need reference models to check VP correctness. Figure 3.1 shows a part of Fig. 1.6 with a *general VP verification* environment in dark gray background. The general VP verification environment requires a *reference model* for *assertions/checkers* to check if the VP testsuite executes the *Design Under Verification* (DUV) as expected. However, the main challenge in this regard is the availability of reference models for verification. When speaking about reference models, it broadly covers approaches such as co-simulation (for instance with Matlab/Simulink), or advanced testbench concepts based on the *Universal Verification Methodology* (UVM), and in the future even more abstract based on *Portable Stimulus Specification* (PSS). Regardless of the specific solution, significant effort is needed to create effective test-stimuli and specify the reference behavior in an executable way. To overcome this problem, we look into *Metamorphic Testing* in this chapter. MT has been first considered in the software domain, and a major advantage of this technique is that no reference model/value is needed, which has to be there in classical software testing. The core idea of MT is to relate consecutive executions of the program under test by the so-called *Metamorphic Relations* (MRs). Hence, the general VP verification environment is extended by several components as shown in green area. They form the basis of MT-based verification approach for AMS VPs as detailed in this chapter.

First, Sect. 3.1 discusses MT-based system-level verification approach. Afterward, Sect. 3.2 presents a novel MT-based verification approach to verify the linear and non-linear behaviors of *Radio Frequency* (RF) amplifiers at the system level. For the class of *Low Noise Amplifiers* (LNAs), 12 high quality MRs are identified. The effectiveness of the proposed MT-based verification approach is demonstrated in an extensive set of experiments on an industrial system-level LNA model without the need of a reference model. This approach has been published in [55].

Furthermore, this chapter goes beyond pure analog systems, i.e., it expands MT to verify AMS systems in Sect. 3.3. As a challenging AMS system, an industrial *Phase-Locked Loop* (PLL) is considered. A set of eight generic MRs is devised.

M. Hassan et al., *Enhanced Virtual Prototyping for Heterogeneous Systems*, https://doi.org/10.1007/978-3-031-05574-4_3

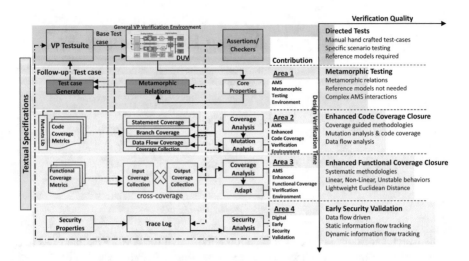

Fig. 3.1 Emerging verification techniques

Theses MRs allow to verify the PLL behavioral at the component level and the system level. Therefore, MRs are created considering analog-to-digital as well as digital-to-digital behavior. The quality and potential of MT for AMS verification are demonstrated using the industrial PLL. This approach has been published in [56].

3.1 MT-Based System-Level Verification Approach

Recently, a new verification perspective has been introduced in the software domain: MT [20, 101, 102], which does not require reference models for verification. Instead of relying on the reference value computed from reference models, MT looks at MRs, i.e., how the inputs and outputs of multiple DUV executions relate.

MRs differ from properties as defined and used in classical verification environments in the AMS domain, such as [15, 32, 35, 62, 78, 80, 92, 97]. For example, consider a DUV that implements a *sum* function for adding two numbers (Fig. 3.2). It takes as input two numbers, 2 and 5, and expected output is 7 (reference value). A MR can be $10 \times sum(2, 5) = sum(10 \times 2, 10 \times 5)$, where the first execution of *sum* (*left hand side* (LHS) of the MR) has the inputs 2 and 5 (termed base test-case), and the second execution (*right hand side* (RHS) of the MR) has the inputs 2×10 and 5×10 (termed follow-up test-case). Instead of verifying what the output of each execution would be, MT only checks if both sides of the MR are equal, i.e., the DUV output of first execution when multiplied by 10 should equal the DUV output of second execution. If the MR is not satisfied, i.e., *LHS $\neq$ RHS*, MT has found a bug. As a consequence, MT does not need a reference model. Furthermore, each MR inherently creates follow-up test-cases using successful base test-cases.

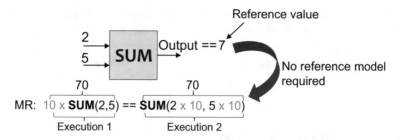

Fig. 3.2 Metamorphic testing example—SUM function

As a consequence, MT can effectively complement regression testing in two ways: (1) generation of follow-up test-cases and (2) the given test-stimuli from regression testing does not need reference models for checking DUV behavior. Employing MT, a large number of real-life faults have been found in complex software where reference models have not been available, see, e.g., [31, 74, 113]. First, we present an overview describing the main components required for MT-based verification approach. Afterward, the different components are briefly explained.

3.1.1 Overview

A high-level block diagram of the proposed MT-based verification approach is shown in Fig. 3.1 (green area). It consists of three major components: (1) test-case generator, (2) metamorphic relations, and (3) core properties of DUV. *Test-case generator* uses the given test-stimuli (base test-cases), i.e., test-stimuli created by *directed testing* and *regression testing*, and MRs to generate follow-up test-cases. The given base test-cases and the follow-up test-cases are input to the DUV to exercise different behaviors of the model. Please note that base test-cases and base test-stimuli will be used interchangeably from this point on. Similarly, follow-up test-cases and follow-up test-stimuli will be used interchangeably from this point on. The *metamorphic relations* block collects DUV outputs for performing relation checks, i.e., comparison of LHS and RHS according to the MRs. The *core properties* component outlines the necessary properties of the DUV that should always be satisfied in a functional DUV. In the next sections, the components of MT-based approach are briefly explained.

3.1.2 Test-Case Generator

As shown in Sect. 3.1, MRs require multiple executions of the DUV with varying inputs. Hence, multiple test-stimuli are required, which build on top of a base test-stimuli. Therefore, the selection of base test-stimuli is of utmost importance because it lays the foundation of the follow-up test-cases. The test-stimuli from the verification plan created during the regular verification process (*directed testing*) are a good candidate as a base test-stimuli. Similarly, test-stimuli from regression testing can also be used.

3.1.3   Metamorphic Relations

The base test-cases and the follow-up test-cases exercise different behaviors of the DUV. Verification of the correct DUV behavior is carried out in the *Metamorphic Relations* block. As motivated earlier, MRs do not need reference models. Hence, this block performs relation checks, i.e., compares the LHS and RHS of MRs. If both the sides of the MR are equal, the MR passes; otherwise, it fails.

3.1.4   Core Properties

The MRs are created using the necessary core properties of a DUV w.r.t. inputs and their corresponding outputs that should always be satisfied. However, while extracting the core properties, the following two things should be considered:

1. Not all core properties of the DUV can be transformed into a MR.
2. MRs are not limited to *equality* relations only. They can have *inequality* relations, or the output of one execution can be a subset of second execution. Hence, the core properties that do not follow an equality relation can also be transformed into a MR.

As reference models are not required to ensure correctness when performing MT, the investment to use the proposed MT-approach is low. However, the potential benefit in design verification is huge as motivated earlier. In the next section, we discuss the MT-approach for RF amplifiers.

3.2   Metamorphic Testing for RF Amplifiers

This section discusses a MT-based verification approach to effectively verify the linear and non-linear behaviors of RF amplifiers at system level. As a representative of RF amplifiers, *Low Noise Amplifiers* (LNAs) are considered. A high quality set of 12 MRs for the class of RF amplifiers is introduced. In an extensive set of experiments on an industrial configurable system-level LNA model, the MT-approach found a serious bug that escaped during the regular verification process. Furthermore, a fault-injection campaign on the industrial LNA is performed. It demonstrates the fault-detection quality of our MT-based verification approach. Consequently, MT-based verification approach successfully verifies the linear and non-linear behaviors of the LNA without the need of a reference model.

Summarizing the main contributions of this section are:

- Novel MT-based verification approach for AMS verification at system level
- A high quality set of MRs (applicable for all variants of RF amplifiers)
- Demonstration of MT-effectiveness on industrial LNA model without the need of reference models

First, Sect. 3.2.1 presents how to transfer the MT principles to the domain of RF amplifiers. Then, Sect. 3.2.2 identifies generic MRs as basis for the overall MT-approach that is introduced in Sect. 3.1.

3.2.1 MT Principle for RF Amplifiers

To leverage MT for verification of RF amplifiers, the central element of MT, i.e., a set of MRs, has to be identified. Recall that a MR is a necessary property of the target function (so in our case an RF amplifier) in relation to multiple inputs and their expected outputs. Linearity of an RF amplifier is one such property where the RF amplifier increases the power level of an input signal without altering the content of the signal. We demonstrate the MT principle for RF amplifier with an example. Let us consider a concrete RF amplifier and its behavior that is defined as $output = 7 \times input$, i.e., the amplifier input is amplified by 7 times (a gain factor of 7). To verify the RF amplifier functionality, the linearity property is converted into the following concrete MR: $3 \times Amp_{vout}(x(t)) = Amp_{vout}(3 \times x(t))$.[1] The MR states that the output voltage of the first execution scaled by a factor of 3 should always be equal to the output voltage of the second execution with an input scaled with a factor of 3. Let us consider this graphically: The base test-stimulus $x(t) = sin(2\pi 5000t)$ is shown in Fig. 3.3a, and the corresponding output signal is shown in Fig. 3.3b, which is 7 times $x(t)$, i.e., $7 \times sin(2\pi 5000t)$. The output signal of the follow-up test-stimulus $3 \times x(t) = 3 \times sin(2\pi 5000t)$ (Fig. 3.3c) is shown in Fig. 3.3d. According to the MR from above, it should hold now that $3 \times Amp_{vout}(x(t))$, i.e., $3\times$ Fig. 3.3b, equals Fig. 3.3d, which is $Amp_{vout}(3 \times x(t))$. This is obviously satisfied here. In case the MR does not hold, the amplifier is termed buggy.

In the next section, we generalize this principle and identify 12 MRs for RF amplifiers as basis for the proposed MT-approach.

---

[1] The concrete amplitude factor of 3 is generalized later.

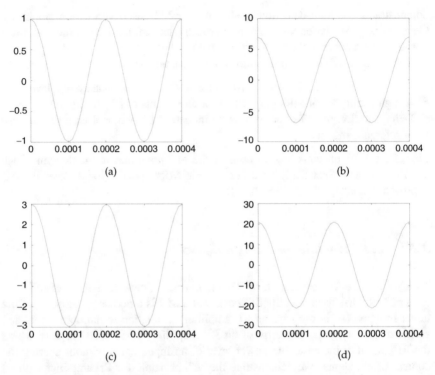

Fig. 3.3 Graphical illustration of MR for amplifier *output* = 7 × *input*. (**a**) Base test-stimulus: $x(t) = sin(2\pi 5000t)$. (**b**) Amplifier output at base test-stimulus. (**c**) The follow-up test-stimulus: $3 \times x(t) = 3 \, sin(2\pi 5000t)$. (**d**) Amplifier output at the follow-up test-stimulus

3.2.2   Identification of Metamorphic Relations

Before the introduction of MRs, the conventions are outlined that will be used for all MRs detailed in the text below. A function $f(x(t))$ is defined in Eq. 3.1 as

$$f(x(t)) = Amp_{char}(x(t)) \tag{3.1}$$

$$x(t) = A \, sin(2\pi F t + \varphi).$$

where Amp_{char} is the amplifier output for a characteristic *char* at an input signal $x(t)$, A is the amplitude, F is the frequency, and φ is the phase of $x(t)$. For future references, $x(t)$ will be used as x interchangeably. *char* can be any one of the following:

$$char = \begin{cases} pout & \text{output power of amplifier (dBm)} \\ gain & \text{gain of amplifier (dB)} \\ vout & \text{output signal voltage (V)} \\ freq & \text{output signal frequency (Hz)} \end{cases}$$

Additionally, the transfer characteristics of amplifiers vary at small-signal levels and high power levels, and as a consequence, the MRs should be developed taking this behavior into account. Hence, some of the identified MRs for both the linear and non-linear operating regions are presented below.

MR1 The output voltage of the RF amplifier scales by the same factor with which the input voltage is scaled. Let x_1 be the base test-case, x_2 is a follow-up test-case, and N is the scaling factor, where $x_2 = N \times x_1$, such that both signals are driving the amplifier in linear region, and then the following should always hold:

$$N \times Amp_{vout}(x_1) = Amp_{vout}(x_2).$$

MR2 The output voltage corresponding to the sum of any two input signals is the sum of the two outputs. If x_1 is a base test-case and x_2 is a follow-up test-case such that both signals drive the amplifier linearly, then the following should always hold

$$Amp_{vout}(x_1 + x_2) = Amp_{vout}(x_1) + Amp_{vout}(x_2).$$

MR3 The output power of the amplifier at x_1 will always be lower than the output power of the same amplifier at x_2. If x_1 is a base test-case and x_2 is a follow-up test-case such that $x_1 < x_2$, and both the signals drive the amplifier in linear region, then the following should always hold:

$$Amp_{pout}(x_1) < Amp_{pout}(x_2).$$

MR4 The output power of the amplifier at x_1 will always be greater than the output power of the same amplifier at x_2. If x_1 is a base test-case and x_2 is a follow-up test-case such that $x_1 > x_2$, and both the signals drive the amplifier in linear region, then the following should always hold:

$$Amp_{pout}(x_1) > Amp_{pout}(x_2).$$

MR5 The amplifier gain should be constant for a range of input signals. If x_1 is a base test-case and x_2 is a follow-up test-case, such that both signals drive the amplifier linearly, then the following should always hold:

$$Amp_{gain}(x_1) = Amp_{gain}(x_2)$$

$$\forall x_1 < x_2, x_1 > x_2.$$

MR6 The output power of amplifier should be constant for a range of input signals. If x_1 is a base test-case and x_2 is a follow-up test-case such that both x_1 and x_2 are very high and the amplifier is operating in saturation, then the following should hold:

$$Amp_{pout}(x_1) = Amp_{pout}(x_2)$$

$$\forall x_1 < x_2, x_1 > x_2.$$

MR7 The amplifier gain should decrease as the input signal power increases, for a range of input signals. If x_1 is a base test-case and x_2 is a follow-up test-case, where x_1 and x_2 are very high and the amplifier is operating in saturation, then the following should hold:

$$Amp_{gain}(x_1) > Amp_{gain}(x_2)$$

$$\forall x_1 < x_2.$$

MR8 All the harmonic distortions of the RF amplifier should shift as the input signal frequency increases. If x_1 is a base test-case with frequency F_1, it results in output harmonics at F_1, $2F_1$, $4F_1$, and the signal x_2 is a follow-up test-case with $F_2 = N \times F_1$, then the following should hold:

$$Amp_{freq}(x_2) = Amp_{freq}(x_1(NF_1)) , \ Amp_{freq}(x_1(2NF_1)) , \ Amp_{freq}(x_1(4NF_1)).$$

MR9 Third-order IMD can cause interference to the desired signal frequencies (F_1 and F_1') because their products are higher in magnitude and close to the desired frequencies. Let x_1 be a two-tone base test-case with signal frequencies F_1 and F_1', where F_1' is slightly lower/higher than F_1. The third-order IMD for x_1 lies at $2F_1 - F_1'$ and $2F_1' - F_1$. Let x_2 be a two-tone follow-up test-case with frequency $F_2 = N \times [F_1 + F_1']$, i.e., the frequencies of x_1 are scaled by a factor of N. Then, the following should hold:

$$Amp_{pout}(x_2) \cong Amp_{pout}(x_1).$$

MR10 The triple-beat test refers to three-tone, third-order IMD and holds a significant place in amplifier verification. Three-tone IMD involves terms of the form $F_1 \pm F_1' \pm F_1''$. If x_1 is a two-tone base test-case with frequencies $F_1 + F_1'$ and the signal x_2 with frequencies $F_1 + F_1' - F_1''$ is a follow-up test-case, then the three-tone third-order IMD is 6 dB higher than two-tone third-order IMD [118], and the following should always hold:

$$Amp_{pout}(x_2) = Amp_{pout}(x_1) + 6dB.$$

MR11 The fundamental principle of *Third-Order Intercept* (TOI) is that for every 1 dB increase in the power of the input tones, the third-order products will increase by 3 dB on the output. If x_1 is a base test-case and $x_2 = x_1 + 1$ dB, $x_3 = x_2 + 1$ dB are follow-up test-cases, such that $Amp_{pout}(x_1)$ gives third-order IMD from x_1 [118], then

$$Amp_{pout}(x_1) + Amp_{pout}(x_2) + Amp_{pout}(x_3) = 3 \times Amp_{pout}(x_1) + 9\,dB.$$

MR12 If x_1 is a base test-case with power P_{x1dBm} and x_2 is a follow-up test-case with power P_{x2dBm}, then the difference in power at input should equal difference in power at output, i.e.,

$$P_{x2dBm} - P_{x1dBm} = Amp_{pout}(x_2) - Amp_{pout}(x_1).$$

Please note that the identified MRs do not represent a complete set of MRs for the amplifiers by any means and more MRs can be identified in principle. The next section presents an extensive set of the experiments on an industrial system-level model using the MT-based approach.

3.2.3 Experimental Evaluation

This section presents the experiments to demonstrate the effectiveness of MT-approach for the verification of linear and non-linear behaviors of RF amplifiers. The experiments use a configurable system-level model of an industrial LNA. Section 3.2.3 provides the details on the LNA model (including the different possible configurations) as well as the experimental setup. Then, Sect. 3.2.3 presents the verification results obtained with the proposed MT-approach. The experiments revealed a serious bug in the LNA using the presented MRs, which escaped during the extensive verification performed earlier. Finally, Sect. 3.2.3 shows the general quality of MT-approach for verification. For this, a fault-injection campaign is performed on the LNA model. This demonstrates that the developed MRs are able to detect all injected faults without the need of a reference model.

LNA Model and Experimental Setup

In general, an LNA amplifies a weak low power input signal without affecting its *Signal-to-Noise Ratio* (SNR) significantly. LNAs can be found in various applications in RF front-ends, e.g., mobile phones, automotive keyless entry devices, wireless LANs, etc. A system-level industrial LNA model (denoted as *I_LNA* in the following text) is considered as a DUV. It is a behavioral model and has been implemented by industry using the well-known concepts from [19, 81]. The specific configuration of the *I_LNA* adheres to the following specifications:

- *Gain* (G) (min., typical, max.) = 16.5 dB, 18.2 dB, 20 dB.
- Input signal amplitude = 0 V to 2 V.
- 1 dB compression point = 30 dBm.
- Output *Third-Order Intercept* (IP3) = 70 dBm.
- Operating frequency = 5 KHz to 20 KHz.
- Input impedance = 50 Ohms.
- Output impedance = 50 Ohms.

The *I_LNA* has been implemented in SystemC-AMS as an abstract static non-linear description as a TDF model. The model comes with a set of test-stimuli created according to the verification plan, i.e., an intensive verification of the linear and non-linear behaviors of the *I_LNA* has already been performed by our industrial partner. Hence, no faults are expected in the model.

MT-Based Verification Results

This experiment uses the test-stimuli of the *I_LNA*, which have been shipped together with the model as mentioned in the previous section. The standard RF specifications of interest have been *gain*, 1*dB compression point*, and the intercept points—*Input Second/Third-Order Intercept* (IIP2/3). As expected, the RF specifications have been verified with the given test-stimuli, and no faulty behavior was observed. The linear and non-linear behaviors were correct. At this point, the MT-approach is employed using the given test-stimuli as the base test-cases. The follow-up test-cases were created with the MT-approach based on the MRs from Sect. 3.2.2. Hence, without any manual effort, our MT-approach immediately found the violation of two MRs: **MR6** and **MR7** failed. The output power and the corresponding gain of *I_LNA* did not follow the *core properties* of the LNA, reflected in these MRs. At higher input voltage, the LNA saturates and the output power should become constant, and the corresponding gain value should decrease in successive test-cases. However, the *I_LNA* did not show this behavior (check **MR6, MR**7 Sect. 3.2.2). Upon close manual inspection of the waveform (Fig. 3.4), a slight overshoot of the gain curve is observed for a short duration before settling to a stable value in the saturation region. This faulty behavior is easy to miss and requires test-cases in a certain range of values to exercise this behavior. However, with the proposed MRs, the follow-up test-cases are computed at no additional cost and even help to debug the model: Based on both failed MRs and the created follow-up test-case, an amplitude increase was observed, which drove the amplifier in the non-linear region. In this region, the underlying approximation algorithm used in *I_LNA* could not handle this non-linearity properly. As a consequence, the gain did not converge, and hence, the relation check (comparison of LHS and RHS according to the MRs) failed. Erroneously, the analog designer has chosen to use *Taylor Series*

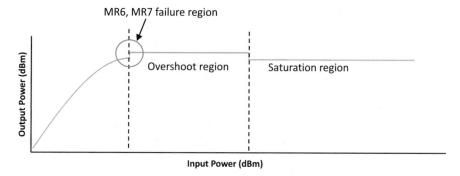

Fig. 3.4 *I_LNA* Output power vs. input power (dBm). Overshoot of output power because of non-linearity approximation

Expansion in the approximation algorithm of *I_LNA* for the non-linearity region and did not consider the occurrence of higher-order polynomials [19]. This bug was discussed with our industrial cooperation partner, and they decided to fix the model accordingly using the concepts from [81]. The model was fixed (denoted *I_LNA-fixed*) and received again for experiments. An excerpt of *I_LNA-fixed* is shown in Listing 3.1. Now, in *I_LNA-fixed*, no further bugs have been found with the MT-based verification approach. The next section analyzes the general quality of MT-approach in detecting bugs.

```
1    ...
2    void lna_base_pb::processing() {
3      double out_temp;
4      if(p.use_iip3_cp1) ()
5        out_temp = s.a*p_in − s.b*pow(p_in,2.0) − s.c_ip3*pow(p_in,3.0);
6      else
7        out_temp = s.a*p_in − s.b*pow(p_in,2.0) − s.c_icp*pow(p_in,3.0);
8
9      double vlim=0.0;
10     // clipping
11     if(p_in < s.in_max && p_in > s.in_min) ()
12       vlim = out_temp;
13     else if(p_in >0.0)
14       vlim = s.a*s.in_max − s.b*pow(s.in_max,2.0) − s.c_ip3*pow(s.in_max,3.0);
15     else
16       vlim = s.a*s.in_min − s.b*pow(s.in_min,2.0) − s.c_ip3*pow(s.in_min,3.0);
17
18     // write outport and signals
19     p_out.write(vlim);
20   }
21   ....
```

Listing 3.1 Excerpt of SystemC-AMS LNA behavior model [81]

Fault-Detection Quality of MT-Based Verification

To show the general quality of MT-approach in detecting faults without the need of a reference model, a fault-injection campaign is performed on the fixed *I_LNA-fixed* as the second set of experiments. The high-level idea is to inject faults in *I_LNA-fixed* to create *mutants* (faulty versions of *I_LNA-fixed*). Then, for these mutants: (a) the shipped test-stimuli are executed and (b) MT-approach is employed.

The fault-injection campaign requires high quality mutants that mimic potential faults. To this end, several faults were injected into the *I_LNA-fixed* SystemC-AMS model. Since the LNA system-level model is in principle a C++ code, therefore, as a fault model, we target common modeling mistakes in the functionality of C++ [3]. The mutants are automatically generated by an *in-house* tool implemented using the *LibTooling* library for *Clang* compiler [58].

The tool generated a total of 175 mutants for the *I_LNA-fixed* model. The proposed MT-based verification approach detected all the 175 mutants, whereas the shipped test-stimuli missed 52 mutants. The results of only 5 test-cases for the experiments are shown in Fig. 3.5 for visual clarity. The *x-axis* shows the developed MRs, and the *y-axis* shows the detected mutants by each MR. The total time it took to simulate one test-case and 175 mutants was approximately 13 min. This time includes compilation, base test-stimuli execution, and follow-up test-stimuli execution. All MRs managed to detect at least one fault, which indicates the different fault-detection quality of the MT-based verification. One concrete *faulty LNA* (FLNA) is discussed to show the effectiveness how MRs detected the fault:

FLNA Consider a mutation where we negate the condition in Line 4 of Listing 3.1, i.e., *if(!p.use_iip3_cp1)*. This injected fault makes the LNA a linear device; hence, it never saturates. When the given test-stimuli are used as input, the LNA behaves linearly as expected and the fault is not detected. The corresponding *gain (dB)* and

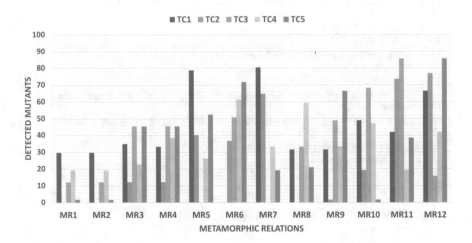

Fig. 3.5 Fault-detection quality of MT-based verification

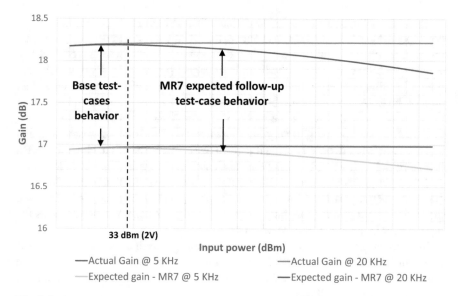

Fig. 3.6 Gain output of FLNA

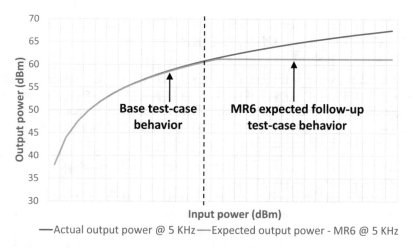

Fig. 3.7 Power output of FLNA

output power (dBm) curves of the LNA are shown in Figs. 3.6 and 3.7, respectively. They have input power (dBm) on the *x-axis* and gain/output power on the *y-axis*. Different color/marker curves represent different frequencies. Simply put, the injected bug cannot be detected using the given test-stimuli.

However, the MT-based verification approach can detect the fault using both the given test-stimuli and the follow-up test-stimuli. The *relation check* verifies the DUV correctness by comparing the sides of MRs. In this case, out of 12 MRs, only **MR6** and **MR7** were able to detect the faulty behavior. Considering **MR7** that looks at the saturation of the LNA, it states that at high input power, i.e., power corresponding to 2 V (33 dBm) and above, the DUV gain should decrease such that the gain at the follow-up test-case should be lower than the base test-case. When we apply a follow-up test-case with an input signal power of 36 dBm, we observe that MR7 fails. The corresponding behavior can be seen in Fig. 3.6 where the *actual gain* of the LNA stays constant (blue lines). The LNA never saturates and passes all the given test-stimuli but fails the **MR7**. Similarly, **MR6** looks at the output power of the LNA corresponding to very high input power. The base test-cases pass, and still the fault is not detected during the given test-stimuli execution. However, **MR6** is able to detect it using the follow-up test-case, as shown in Fig. 3.7. **MR6** looks at the output power of the amplifier in saturation and expects that it should remain constant when the base test-case and the follow-up test-case are applied. However, the actual output power (blue line) as shown in Fig. 3.7 keeps on increasing as the input power increases; hence, **MR6** fails.

To summarize the experiments, the MT-based approach effectively verifies the linear and non-linear behaviors of the RF amplifiers without the need of reference models. In the next section, the MT-based approach is extended toward a complex AMS system for its verification.

3.3 Metamorphic Testing for PLLs

Section 3.2 presented a successful application of MT to the verification of RF amplifiers at the system level. However, this is clearly not sufficient as the true complexity stems from AMS systems. This section goes beyond pure analog systems, i.e., it expands MT to verify AMS systems. The following contributions are made in the following sections:

1. An industrial *Phase-Locked Loop* (PLL) AMS system is considered, and a set of eight generic MRs is devised.
2. It is demonstrated that the identified MRs allow to verify the PLL behavior at the component level and the system level. Therefore, the MRs consider analog-to-digital, digital-to-analog, as well as digital-to-digital behavior.
3. Besides successful verification of a broad spectrum of tests, the identified MRs can be easily used to derive follow-up test-cases at different levels and hence improve the verification. Please note that these test-cases and the MRs can be reused at lower abstraction levels.
4. A critical bug in the industrial PLL model was revealed, which clearly demonstrates the quality and potential of MT for AMS verification.

The next section introduces the industrial PLL. Afterward, it briefly introduces the MT principle for mixed-signal interactions. Then, we identify the generic MRs for PLLs. At the end, experimental results are discussed.

3.3.1 Phase-Locked Loop

A PLL operates by comparing an input frequency with the system's clock frequency and subsequently adjusting its output to match the input. It comprises a *Phase Frequency Detector* (PFD), *Charge Pump* (CP), *Loop Filter* (LF), *Voltage Controlled Oscillator* (VCO), and a *Frequency Divider* (FD). The PFD compares the phase of input signals and accordingly sends two signals UP/DN to CP. As a result, the CP generates pulses of positive/negative currents. This current goes to LF that generates a control voltage signal. The control voltage is applied to the VCO that generates the output frequency. The FD takes the output frequency signal and divides it by a factor N. The resulting signal goes back to PFD. PLLs are widely used in carrier recovery, clock recovery, frequency synthesis, clock synchronization, etc. A configurable system-level model of PLL [25, 29] provided by our industrial partner is used.

The high-level PLL block diagram is shown in Fig. 3.8[2] and has the following specifications:

- Free running frequency $(F_{osc}) = 2.39\,\text{GHz}$
- $vdd = 3.3\,\text{V}$, $vcm = 1.65\,\text{V}$
- CP current_up/dn $= 100\,\mu\text{A}$
- VCO Gain $= 36.36e^6$
- Frequency Divider $N = 2450$

The model is implemented in SystemC AMS using *Timed Data Flow* (TDF) and *Electrical Linear Network* (ELN) *Models of Computation* (MoC) for different building blocks. The model also uses *Discreet Event* (DE) simulation. The simulations are carried out using the commercial tool COSIDE [115]. Test-stimuli are provided with the model to verify the functional correctness.

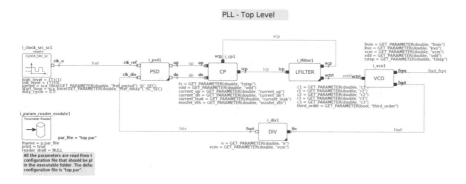

Fig. 3.8 PLL top level diagram

---

[2] PFD is written as PSD in the provided model as shown.

3.3.2   MT Principle for Mixed-Signal Interactions

Mixed-signal interactions require understanding of both the analog–digital inputs and the corresponding digital–analog outputs w.r.t. the core properties of the DUV. One such property highlighting these interactions is related to CP in PLL where the digital inputs vary the analog current on the output of the CP. More concretely, let us consider a CP with the following behavior:

$$icp = \begin{cases} +100e^{-6} & UP = 1, \ DN = 0 \\ -100e^{-6} & UP = 0, \ DN = 1 \end{cases},$$

i.e., CP checks its input and generates a positive/negative pulse of $100e^{-6}$. To verify this, a concrete MR can be: $icp[CP(1, 0)] = -icp[CP(0, 1)]$. The MR states that the output current icp of the first execution (LHS)[3] should always be equal to the negative of output current icp of the second execution (RHS).[4] Let us consider this graphically: The base test-case $CP(1, 0)$ is shown in Fig. 3.9 (top waveform) from time $t = 0.0$ to $0.5\,\mu s$, and the corresponding icp is shown in Fig. 3.9 (bottom waveform), which is $+100e^{-6}$ A. The follow-up test-case (inverted inputs) $CP(0, 1)$ is shown in Fig. 3.9 (middle waveform) from time $t = 0.5$ to $1.0\,\mu s$, and the corresponding icp is shown in Fig. 3.9 (bottom waveform), which is $-100e^{-6}$ A. According to the MR above, it should hold that $icp[CP(1, 0)]$, i.e., $+100e^{-6}$, equals $-(-100e^{-6})$, which is the icp when inputs are inverted, i.e., $icp[CP(0, 1)]$. This is obviously satisfied here. In case the MR does not hold, the CP is termed buggy.

The next section generalizes this principle and identifies eight MRs for PLL and few of its components.

3.3.3   Identification of MRs for PLLs

This section describes eight high quality generic MRs identified from the core properties of PLLs and its components. The analog-to-digital, digital-to-analog, and digital-to-digital interactions are considered for the MRs. As a convention, F_{osc} is the free running frequency of the PLL when either there is no input frequency or the input frequency is out of capture range. F_{DIV} is the PLL output frequency F_o divided by N, i.e., $F_{DIV} = \frac{F_o}{N}$.

MR1 The digital output UP of PFD is $True$ if the analog input frequency F_1 is higher than F_2. Similarly, the digital output DN of PFD is $True$ if the analog input frequency F_1 is lower than F_2. Hence, the following should always be satisfied

---

[3] LHS: left hand side.

[4] RHS: right hand side.

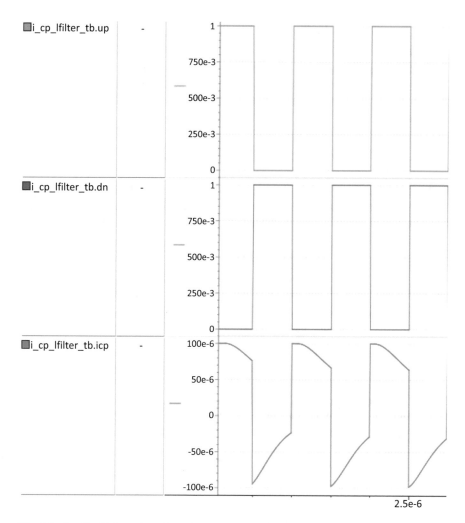

Fig. 3.9 Graphical illustration of MR for charge pump (CP)

across 4 executions of PFD with varying input frequencies F_1 and F_2:

$$UP[PFD(F_1, F_2)] + UP[PFD(F_2, F_1)] = DN[PFD(F_1, F_2)]$$
$$+ DN[PFD(F_2, F_1)].$$

MR2 The CP generates pulses of positive or negative currents based on its digital inputs. Let $X_1 = IN_1, IN_2$ be the base test-case, and let $X_2 = IN_2, IN_1$ be the follow-up test-case; then, the following relation should always be satisfied over 2 executions of the CP:

$$ICP[CP(X_1)] = -ICP[CP(X_2)].$$

MR3 When the PLL is not in locked state, F_o is always at free running frequency F_{osc}. Let F_1 be the base test-case and F_2 be the follow-up test-case, such that both F_1 and F_2 are not in lock range; then the following should always hold across 2 executions:

$$F_o[PLL(F_1)] - F_o[PLL(F_2)] = 0.$$

MR4 In PLL locked state, F_{DIV} scales by the same constant with which the input frequency is scaled. Let F_1 be the base test-case and $F_2 = C \times F_1$ be a follow-up test-case where C is the scaling constant; then the following should always hold:

$$C \times F_{DIV}[PLL(F_1)] = F_{DIV}[PLL(F_2)].$$

MR5 If PLL is not locked, F_o equals F_{osc}. Let F_1 be the base test-case when the PLL is in locked state such that $F_1 > Fosc$, and F_2 be a follow-up test-case such that F_2 is very high and outside PLL lock range; then the following should always hold:

$$F_{DIV}[PLL(F_1)] > F_{DIV}[PLL(F_2)].$$

MR6 If PLL is not locked, F_o equals F_{osc}. Let F_1 be the base test-case such that $F_1 < F_{osc}$, and F_2 be the follow-up test-case such that F_2 is outside lock range (very high or very low); then the following should always hold:

$$F_{DIV}[PLL(F_1)] < F_{DIV}[PLL(F_2)].$$

MR7 The PLL stays in the locked state (indicated by *Lock Detector* (LD)) if the input frequency is varied inside the lock range. Let F_1 be the base test-case, and F_2 and F_3 be the follow-up test-cases, such that all the frequencies keep the PLL in locked state, then the following should always hold:

$$LD[PLL(F_1)] \& LD[PLL(F_2)] \& LD[PLL(F_3)] = 1.$$

MR8 F_o synchronizes to a new frequency within the lock range in a single beatnote [38]. Let F_1 be the base test-case and F_2 be the follow-up test-case such that the PLL is in locked state at both frequencies; then the following should always hold:

$$LockTime[PLL(F_1)] = LockTime[PLL(F_2)].$$

3.3.4 Experiments

This section presents the experiments to demonstrate the quality and potential of MT for system-level AMS verification.

Overview

As already mentioned in the introduction, a configurable industrial system-level PLL model is used, which is provided by our industrial collaboration partner. The details and specifications of the PLL have been already described in Sect. 3.3.1. The 8 MRs devised in Sect. 3.3.3 are used to verify the PLL behavior at component level and the system level. A critical bug was found in the PLL design using the introduced MRs, which has escaped during the extensive verification. In the following, more details on the proposed MT-based approach are provided.

MT-Based Verification of PLL

The test-cases shipped with the model are used as test-stimuli. As expected, the simulations for the set of shipped test-stimuli pass. As a next step, the MT-approach is employed using the given test-stimuli as the base test-cases. The MRs from Sect. 3.3.3 were used to create the follow-up test-cases. Out of these, 10% follow-up test-cases covered analog-to-digital behavior at the component level, and 10% follow-up test-cases covered analog-to-digital behavior at the system level. Furthermore, 10% covered digital-to-analog behavior at the component level, and the remaining 70% follow-up test-cases covered the digital-to-digital behavior at the system level.

Running the MT-approach with the proposed MRs resulted in a simulation failure. More precisely, **MR4** was not satisfied: The constant factor $C = 1.01$ increased the input frequency (F_1) of the PLL (RHS of MR4 with $F_2 = C \times F_1$), from 1 MHz to 1.01 MHz, and it was expected that the divided frequency F_{DIV} will also increase by the same factor C (LHS of MR4). However, that was not the case.

Upon close inspection of the waveforms of F_{REF} and F_{DIV} (inputs of PFD) and *Fast Fourier Transform* (FFT) of F_{REF} and F_{DIV} revealed that the PLL was locking to a different very low frequency of 50 KHz instead of 1.01 MHz. The faulty behavior is shown in Fig. 3.10 where F_{REF} is a high frequency signal of 1.01 MHz (top waveform) and F_{DIV} is unexpectedly low frequency signal of 50 KHz (bottom waveform). The FFT of F_{REF} and F_{DIV} signals is shown in Fig. 3.11 (top waveform—F_{REF}, bottom waveform—F_{DIV}). The FFT of F_{REF} shows a peak at 1.01 MHz, and the FFT of F_{DIV} shows multiple peaks at various frequencies with the strongest peak at 50 KHz. Upon further investigation, we observed the **dead-zone** effect, i.e., a **dead zone** was occurring in the output behavior of the PFD. A **dead zone** occurs when the PLL loop does not respond to small phase errors

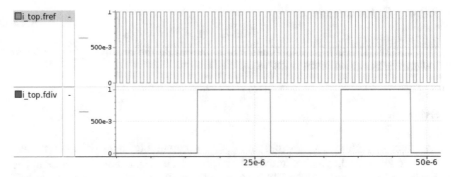

Fig. 3.10 PLL faulty behavior—dead-zone effect revealed by MR4

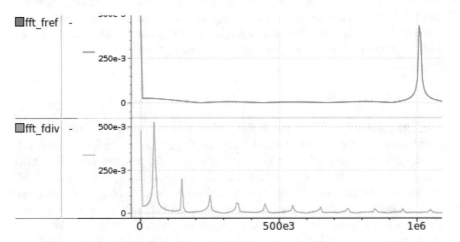

Fig. 3.11 FFT of PLL faulty behavior

between F_{REF} and F_{DIV}. As a result, the output of CP is modulated by a signal that is a sub-harmonic of the PFD input reference frequency F_{REF}. Since this could be a low frequency signal, it would not be attenuated by LF [5]. Looking into the design of PFD revealed that there was no delay element between the *AND* gate and the *reset* pins of the flip-flops. The design without any delay element is shown in Fig. 3.12. The delay element between the output of *AND* gate and the *reset* inputs of flip-flops ensures that dead-zone effect does not happen. After insertion of the delay element, we observed the correct output behavior of the PLL and **MR4** was satisfied (cf. Fig. 3.13). The F_{REF} and F_{DIV} signals can be observed at same frequencies.

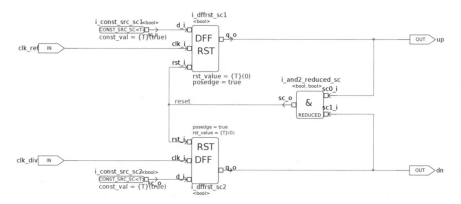

Fig. 3.12 Phase frequency detector (PFD) without a delay element

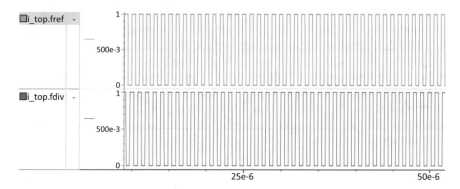

Fig. 3.13 PLL behavior after addition of delay element in PFD

3.4 Summary

We have introduced a novel MT-approach for the verification of linear and non-linear behaviors of RF amplifiers at the system level. We leveraged the MT principles and identified a set of 12 high quality MRs. In an extensive set of experiments on an industrial system-level LNA model, we have demonstrated the ability to find non-trivial bugs without the need of reference models. Furthermore, we broadened the MT-approach to verify complex AMS systems, in particular an industrial PLL. We identified 8 high quality generic MRs to verify the PLL behavior at the component level and at system level that encompasses analog-to-digital, digital-to-analog, and digital-to-digital behaviors. During experiments, we found a critical bug in the PLL that demonstrates the quality and potential of MT for AMS verification.

Additionally, at the end, we discuss some observations w.r.t. the proposed MT-approach as well as limitations and possible improvements. As can be seen from the experiments (cf. Fig. 3.5), it is not guaranteed that all the faults will be detected by a single MR. Moreover, there are MRs that detect a high number of faults, and others that detect a very small amount, sometimes even 0. This is mainly because of two reasons: (1) nature of the fault and (2) how the fault is related to an MR. Also, this has been reported for MT in the software domain, see for instance [20, 21]. As a solution, these works proposed to develop and use many MRs. Moreover, when comparing MT to the traditional verification techniques, there is of course some effort for deriving MRs. However, this is a one-time effort, and in particular for the systems as considered here a lot of general MRs can be defined for one class of designs. MRs need to be identified as part of the planning and design phase and included in the DUV specification. Here our suggestion is to work on collections of MRs and share them in our community. Furthermore, one of the applications for MRs is to automate regression testing as motivated in the introduction since they can be checked as to whether they hold in different versions of the DUV.

For future work, we plan to investigate the following directions: (1) validate the MT-approach at SPICE-level and (2) apply the MT-approach to other classes of AMS designs. Another very interesting research direction is to devise solutions for determining the completeness of a set of MRs. In the context of regression testing, test-case coverage data is available (either for the previous versions or for the current version of the DUV). We plan to explore such scenarios where MRs can be selected or prioritized directly using the available coverage data. In addition to MRs, the initial and follow-up test-cases may also be prioritized by globally considering the most dissimilar test-cases.

Chapter 4
AMS Enhanced Code Coverage Verification Environment

This chapter discusses novel code coverage closure methodologies to enhance the verification quality of modern VP verification flow. Since a VP is in essence a software model, simulation-based verification for VPs is actually very similar to software testing, and therefore, techniques from this domain can be borrowed to ensure a high quality of verification results. These enable HW/SW co-design and verification very early in the design flow, and in particular, the approaches of *Software Driven Verification* (SDV) [88] and *Data Flow Testing*(DFT) [82, 98] for SystemC/AMS. In this regard, the *general VP verification* environment (dark gray area) in Fig. 4.1 is enhanced by several components as shown in blue area. They form the basis of verification methodologies for AMS VPs as detailed in this chapter. We start in Sect. 4.1 with a novel quality-driven methodology based on *mutation analysis*. By elevating the main concepts of mutation based qualification to the context of SDV, our methodology is capable to detect serious quality issues in the SW tests. At its heart is a novel consistency analysis that measures the coverage of the IP in HW/SW co-simulation in a lightweight fashion and relates this coverage to the SW test results to provide clear feedback on how to further improve the quality of tests. We provide two case studies on real-world VPs and SW tests to demonstrate the applicability and efficacy of our methodology. This methodology has been published in [47].

Furthermore, in Sect. 4.2, we discuss *Data Flow Testing* (DFT) for SystemC VPs. Our contribution is twofold: First, we develop a set of SystemC specific coverage criteria for data flow testing. This requires considering the SystemC semantics of using non-preemptive thread scheduling with shared memory communication and event-based synchronization. Second, we explain how to automatically compute the data flow coverage result for a given VP using a combination of static and dynamic analysis techniques. The coverage result provides clear suggestions for the testing engineer to add new test-cases in order to improve the coverage result. Our experimental results on real-world VPs demonstrate the applicability and efficacy

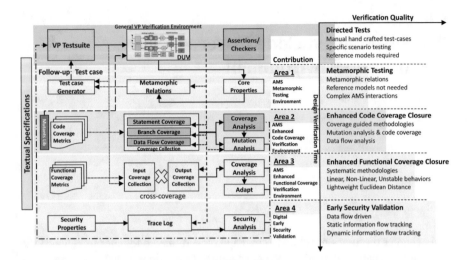

Fig. 4.1 Enhanced structural coverage verification

of our analysis approach and the SystemC specific coverage criteria to improve the testsuite. This approach has been published in [60].

Later, Sect. 4.3 introduces a DFT approach for SystemC-AMS TDF models based on two major contributions: First, we develop a set of SystemC-AMS TDF models-specific coverage criteria for DFT. This requires considering the SystemC-AMS semantics of signal flow. Second, we explain how to automatically compute the data flow coverage result for given TDF models using a combination of static and dynamic analysis techniques. Our experimental results on real-world AMS VPs demonstrate the applicability and efficacy of our approach. This approach has been published in [57].

4.1 Software Driven Verification for IP Integration

As motivated, the much earlier availability and the significantly faster simulation speed in comparison to *Register Transfer Level* (RTL) are among the main benefits of SystemC-based VPs. These enable HW/SW co-design and verification very early in the design flow, and in particular, the approach of *Software Driven Verification* (SDV) proposed in [88]. Essentially, software tests are developed for functional verification of the (new) integrated IP blocks and the HW/SW integration. The tests are typically written in C and run on a processor core of the VP. The key benefit of SDV is that the tests can be reused along all following design phases, i.e., in RTL simulation, emulation, FPGA prototyping, and even the silicon. This is very valuable as IP integration is becoming more and more a bottleneck for today's high-performance SOCs that typically include multiple processor cores and hundreds of IP blocks.

To reap the most benefit from SDV, the quality of the software tests is crucial, as low quality software tests could miss serious integration issues. Hence, in this section, we discuss a novel guided approach to evaluate and improve software tests developed for integration verification of an IP block. It is based on mutation analysis. In the software testing community, mutation analysis [26, 30, 52, 75] has been considered for decades as a fault-based technique. Essentially, it is checked whether the tests are capable of detecting (killing) the deviating behavior of a syntactically correct but modified program (a mutant). The ideas have also been transferred to hardware verification [53, 58, 76, 90, 105, 112] and system-level verification [16, 17, 84, 104]. In this context, the three main tasks of qualification are distinguished: (1) activate, i.e., stimuli have to be provided to activate the mutation; (2) propagate, i.e., the effect of the mutation has to propagate to an observable point; and (3) detect, i.e., the testbench must detect functional mismatches between the original design and the mutated one. However, these qualification tasks give very little information about the nature of mismatches in the compared designs. As a consequence, this book makes a twofold contribution to enable qualification of software tests for verification of IP integration. First, we define the main tasks activation, propagation, and detection in the context of SystemC VP-based IP integration. Building on that, our main contribution goes one step further to provide a complete methodology to *guide* the verification engineer in improving the software test quality. If the mutation in the IP block is not killed by the software tests, the engineer wants to know the reason and improve the tests. For this problem, we propose a novel *consistency analysis* that relates the mutation results with the coverage results of the original (not mutated) IP block verification and provides a guided solution: If they are inconsistent, the methodology gives clear hints when and for which mutants to add more tests and when to use a more powerful coverage model. Following the proposed methodology, a big jump in the quality of the SW testsuite can be achieved in consecutive iterations while using different variants of well-known code coverage models that can be very easily measured. In Sect. 4.1.1, the methodology is introduced.

4.1.1 SW Test Qualification Methodology

In this section, we present the proposed methodology that qualifies SW tests developed for verification of IP integration in VPs with the help of our guidance mechanism. At first, the setting when verifying IP integration is described. Next, the core of the proposed methodology is introduced, i.e., the consistency analysis of coverage measurement and software test result w.r.t. a mutation. Then, the overall methodology is presented. Finally, easy-to-grasp examples are provided to demonstrate the different steps of the methodology.

Setting of IP Integration Verification

In a SDV environment with the task of verifying the integration of new IP, the test creator typically writes a sequence of tests that form the testsuite. These tests interact step-by-step with the IP block, and they are self-checking, i.e., the results of interactions with the IP are checked within each test, e.g., by using C assertions. Ideally, the testsuite should examine the IP thoroughly; otherwise, integration issues could be missed.

As a prerequisite for our methodology, we assume that the tests already achieve a high statement coverage of the IP block. We believe this assumption is fair due to the following reasons. In practice, very often statement coverage of the IP block is measured to ensure that each statement has been exercised, at least by one test. High statement coverage is a positive indicator of the quality of the tests.

It can be achieved quickly by taking the IP block without any mutation and writing a testsuite sufficient to trigger a higher number of statements and branches. By not focusing on any particular area rather going throughout the IP features briefly can prove helpful in maintaining high testing productivity. The strategy is not to have 100% coverage initially, but to have maximum coverage with minimum efforts. If the coverage is low initially, more tests should be added by the testsuite creator. Metamorphic relations can be leveraged in this regard [55, 56] as explained in Chap. 3.

However, statement/branch coverage has severe limitations regarding whether the desired behavior has really been checked. Furthermore, there is typically a point of diminishing returns, i.e., after a high coverage, for example 90%, is achieved, it is very difficult to increase it further. When that happens, the effort should be shifted to a more sophisticated (but still lightweight) qualification methodology. We propose such a methodology for software test qualification. But before we present the overall methodology, we introduce the core of our methodology—consistency analysis—in the following section.

Consistency Analysis

Let us just for a moment assume a single mutation is considered only. Then, the consistency analysis includes the following 4 main steps:

1. Generate coverage report for the original IP running current testsuite.
2. Mutate IP using a fault model.
3. Generate coverage report for mutated IP.
4. Analyze consistency of coverage results and software test result.

In the subsequent sections, we detail the major aspects of each step. Furthermore, we describe the relation to the main qualification tasks (activate, propagate, detect).

Fault Model—Mutation of IP Block When mutating the new integrated TLM IP block, mutations are only performed in its SystemC/C++ code that has been

Table 4.1 Summary of SystemC specific mutation operators

Operator	Original	Mutant
Modify	wait (200)	wait (200/2)
Remove	wait (200)	–
Replace	wait (200)	wait ()
Exchange	trywait ()	wait ()
	wait (200)	notify ()
	wait (event1)	wait (event2)

marked as covered in the code coverage report, i.e., mutations will never be done in dead code (such mutations cannot be activated; hence, their simulation is a waste of time). Essentially, this gives us the *activation* of the mutation since we know based on the coverage that the mutated statement is reachable by at least one test. At this point, the importance of the prerequisite for high code coverage can be emphasized. Because otherwise, mutations could only be applied to a small portion of code limiting its effectiveness significantly.

Since we are "looking" from the software test perspective, mutations that affect the functional behavior of the IP block are the most interesting. Mutations that affect the TLM commutation, for example, modifying register addresses or holding off responses, are for the most part detected by simple checks that are present in SW tests (e.g., write some value to a register, and then check if a read from the register returns the same value). Furthermore, restricting mutation operators to a small number of really relevant classes reduces the overall mutation effort. Therefore, as a fault model, we target common modeling mistakes in the functionality of a SystemC TLM IP. These include both the sequential and concurrent aspects. For sequential modeling faults, we adopt the comprehensive set of mutation operators as proposed in [3], where 77 C/C++ mutation operators are explained. The mutation operators are categorized into four domains: statement mutations, operator mutations, variable mutations, and constant mutations. They are primarily based on the competent programmer hypothesis, i.e., faults are syntactically small and only few keystrokes away from the original program. Since the IP has already passed the initial SW testsuite with a high statement coverage, this hypothesis is also plausible in our setting. Furthermore, our set of mutation operators is extended by SystemC specific mutation operators as proposed in [103]. These operators target TLM communication and synchronization with a particular focus on concurrency constructs. They are summarized in Table 4.1 (for the details, we refer to reader to [103]).

Coverage of Mutated IP Measuring code coverage of the mutated IP is straight-forward. Note that code coverage allows to observe the *propagation*. This is similar to but simpler than *Control Flow Graph* (CFG)/*Data Flow Graph* (DFG) based propagation detection, which requires more complex source code analyses (see Sect. 4.2 for more details). In a perfect setting of course, a propagation monitor would be used, which checks at the boundary that the mutation leads to

Table 4.2 Consistency analysis results for a mutation

Category	Coverage (propagate)	Result of SW test (detect)	Consistent	Interpretation
C1	Fluctuate	Fail	Yes	Adequate test
C2	Fluctuate	Pass	No	Weak detection by SW test
C3	Stable	Fail	No	Propagation path missing
C4	Stable	Pass	Yes	Propagation/detection problem

a difference. However, the definition of boundary is not obvious. Also, for such a propagation monitor, detailed knowledge of the IP would be required, e.g., to specify the corresponding *SystemVerilog Assertions* (SVA) properties. Furthermore, the "natural" boundaries for the monitor might be only observable at the IP level but not from the perspective of SDV; hence, additional effort might be required to lift these to the software level. Before such effort becomes necessary, we propose to consider code coverage as a lightweight alternative for observing propagation.

Consistency Analysis—Comparison of Coverage Result and Software Test Result In this section, first the principles of the consistency analysis are introduced. Then, it is shown how to measure the quality of the SW tests in the form of a consistency score. Different results are possible for an injected mutation when analyzing the consistency of the coverage result and the SW test result. The possible results are summarized in Table 4.2. Column *Category* assigns a number to each of the four possible categories. The second column *Coverage* lists whether the coverage of the original IP block when running the testsuite (Step 1 of the overall consistency analysis as described earlier) in comparison with the coverage results (Step 3) changes or not. In case of a difference, this is labeled as *fluctuate*; if it remains unchanged, it is labeled as *stable*. The column *SW Test Results* shows whether the execution of the software testsuite resulted in *fail* or *pass*. The next column *Consistent* defines whether the comparison outcome of the coverage result and the software result is consistent or not. In the last column *Interpretation*, a short intuitive explanation is given. A more detailed explanation is provided in the following:

C1: If the coverage fluctuates and the SW test fails, their behavior is consistent since the propagation has been recognized by code coverage, and due to the mutation, at least one test fails as expected. As a consequence, for the current mutation, the tests are adequate.

C2: If the coverage fluctuates and the SW test passes, the situation is inconsistent. The propagation is recognized (manifesting in the change of coverage), but the software tests unexpectedly pass. Hence, the detection is weak, meaning that a test should be added to improve the testsuite.

C3: If the coverage is stable and the SW test fails, again the situation is inconsistent. The reason for inconsistency is that code coverage could not recognize the propagation path. This inconsistency is not considered harmful because the mutant still gets killed. Hence, a C3 mutant does not require an action on its own. Instead, C4 category is consulted.

C4: If the coverage is stable and the SW test passes, the situation is consistent. However, the software tests should not pass when performing a mutation. Different reasons are possible for this scenario, so we have to deal with both the propagation and detection problem. We know that the propagation is problematic, especially if C3 mutants are also present. A potential solution is the use of a stronger coverage metric, for instance branch coverage.

The quality of SW testsuite after consistency analysis can be measured with the help of a *consistency score* in a similar manner to the established *mutation score* or *mutation adequacy* [18, 48] as follows:

$$CS = \frac{\#C1}{\#C1 + \#C2 + \#C3 + \#C4} \tag{4.1}$$

In Eq. 4.1, Cx is the total number of mutants in category Cx (with $x \in \{1, 2, 3, 4\}$). Please note that in the numerator only C1 is used and not the sum of C1 and C3. The reason is that only category C1 is consistent *and* has positive interpretation, i.e., the testsuite is adequate.

Based on the introduced consistency analysis for a mutation, we present our methodology in the next section.

Overall SW Qualification Methodology

In Fig. 4.2, the overall methodology is depicted. It starts at the top of the figure with the mutation database containing all possible mutations for the current IP block. The overall goal of the proposed methodology is to finally bring all mutants into the category C1. Generally, several iterations might be needed to achieve this goal. In the first iteration, for each mutant from the database, the current SW tests are executed, and the consistency analysis is performed. Depending on the returned category, different actions need to be taken to update the database. In case of C1, the testsuite is adequate for the mutant, so this mutant is removed. Otherwise, the consistency result for the mutant is saved.

After the first iteration (i.e., no mutant left), the updated database is analyzed. If it is empty, which means every mutant has been in category C1 and has therefore been removed, we are done. If at least one mutant of category C2 can be found, new SW tests must be added to kill the mutant, then a new iteration is started. The last possible outcome of the analysis is that the database contains only mutants of categories C3 and C4. If only C3 mutants are alive, the verification engineer can ignore them and consider them as killed (recall the SW test already failed for category C3). But if both categories C3 and C4 are present, then the coverage model should be revised to increase the resolution for propagation. After this, a new iteration is started.

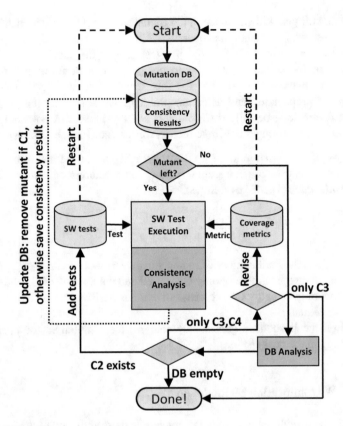

Fig. 4.2 SW qualification methodology

Comparison to Classical Mutation Based Qualification

In comparison to classical mutation based qualification technique, our methodology guides the verification engineer in the correct direction. By limiting the effort of writing new tests to mutants of category C2 only, our methodology ensures that a test is not being written for an equivalent mutant (trying to do so would lead to waste of time and resources). In contrast, the classical technique only gives information on the mutants killed and mutants alive. No information is provided on the nature of mutants. The guidance resulting from the categorization of each mutant through consistency analysis can save a lot of effort and time.

In the next sections, we demonstrate the core of our methodology—consistency analysis—for a simple example, and also two case studies for real-world examples. They also show how the guidance is provided to make the process easier.

```
1    void maxIP::find_max() {            19              else
2      uint32_t a, b, c, d, max;         20                max = d;
3      while (1) {                       21           else
4        wait(e_signal);                 22             if ((b > c) && (b > d))
5        a = r[SRC_A] ;                   23               max = b;
6        b = r[SRC_B] ;                   24             else if ((b > c) && (b < d))
7        c = r[SRC_C] ;                   25               max = d;
8        d = r[SRC_D] ;                   26             else if ((b < c) && (b > d))
9        if (a > b)                       27               max = c;
10         if ((a > c) && (a > d) )        28             else
11           max = a;                      29               if (c > d)
12         else if ((a > c) && (a < d))    30                 max = c;
13           max = d;                      31               else
14         else if ((a < c) && (a > d))    32                 max = d;
15           max = c;                      33         r[MAX_VALUE] = max;
16         else                            34       } // end while
17           if (c > d)                    35     } // end find_max()
18             max = c;
```

Listing 4.1   Code excerpt of IP block maxIP SC_THREAD *find_max*

4.1.2   Consistency Demonstration Example

We use a compact example to demonstrate the ingredients of the methodology.

IP Block Basic Information

A code excerpt of a complete SystemC TLM model of the IP block *maxIP* is shown in Listing 4.1. The main functionality of this IP is implemented in the SC_THREAD *find_max*. It receives four inputs from its registers (*r[SRC_A]*, *r[SRC_B]*, *r[SRC_C]*, *r[SRC_D]*), finds the maximum value among them, and writes this value back into the register *r[MAX_VALUE]*.

SW Tests

Listing 4.2 shows the SW tests for *maxIP*. As can be seen, in each test, four integer values are written into the memory addresses of the registers of *maxIP* (see e.g. Lines 12–15), and then the computed maximum value is read back and compared to the expected value. If the maximum value is correct, the test generates a *success* message (e.g., Line 16 with argument 1 since it is the first test); otherwise, a *fail* message is generated (e.g., Line 17 again with argument 1 indicating the test number).

```
1   struct maxIP {
2     volatile unsigned int SRC_A; /* 0x00 */
3       volatile unsigned int SRC_B; /* 0x04 */
4       volatile unsigned int SRC_C; /* 0x08 */
5       volatile unsigned int SRC_D; /* 0x0C */
6       volatile unsigned int MAX_VALUE; /* 0x10 */
7   };
8
9   void SW_Test(int addr) {
10    struct maxIP *ipBlock = (struct maxIP *) addr;
11    /* Test 1 */
12    ipBlock->SRC_A = 5;
13    ipBlock->SRC_B = 1;
14    ipBlock->SRC_C = 8;
15    ipBlock->SRC_D = 4;
16    if (ipBlock->MAX_VALUE == 8) success(1);
17    else fail(1);
18    /* Test 2 */
19    ipBlock->SRC_A = 5;
20    ipBlock->SRC_B = 1;
21    ipBlock->SRC_C = 3;
22    ipBlock->SRC_D = 6;
23    if (ipBlock->MAX_VALUE == 6) success(2);
24    else fail(2);
25  }
```

Listing 4.2 Consistency example: SW Test

Coverage of SW Tests

The coverage results for the SW tests of Listing 4.2 are depicted in Fig. 4.3 and can be interpreted using Table 4.3. Essentially, each line of Fig. 4.3 consists of three parts: Branch Coverage : Line Coverage : src expression. Let us start with line coverage: As can be seen on the left side of Table 4.3, red color is used to show that the expression has not been hit during execution, and otherwise the color is blue. Moreover, the number of executions is shown as second part in each line of Fig. 4.3.

For branch coverage, only the source code lines with conditions are relevant. Each condition of a branch in Fig. 4.3 has a corresponding [T F] pair (meaning [True False]) shown on the left of each code line. Note that T and F are replaced with the symbols shown in Table 4.3 in column symbol. As can be seen for instance in Line 186 of Fig. 4.3, it has only one pair, and Line 187 has two pairs, respectively. When the SW tests evaluate a condition with true, the T is replaced

Table 4.3 Consistency example: coverage legend

Branch coverage			
Line coverage			
Color	Line	Symbol	Description
Blue	Hit	+	Taken
Red	Not hit	–	Not taken
		#	Not executed

```
180        Branch      :   Line   :
181  [T  F][ T  F ]:          2 :    a = r[SRC_A] ;
182               :          2 :    b = r[SRC_B] ;
183               :          2 :    c = r[SRC_C] ;
184               :          2 :    d = r[SRC_D] ;
185
186        [ + - ]:          2 :    if (a > b)
187  [ + + ][ - + ]:          2 :        if ((a > c) && (a > d) )
188               :          0 :            max = a;
189  [ + + ][ + - ]:          2 :        else if ((a > c) && (a < d)) //MC4 - replace && with ||
190               :          1 :            max = d;
191  [ + - ][ + - ]:          1 :        else if (((a < c) && (a > d))) //MC2 - negate expression
192               :          1 :            max = c;
193               :            :        else
194        [ # # ]:          0 :            if (c > d)
195               :          0 :                max = c;
196               :            :            else
197               :          2 :                max = d;
198               :            :    else
199  [ # # ][ # # ]:          0 :        if ((b > c) && (b > d))
200               :          0 :            max = b;
201  [ # # ][ # # ]:          0 :        else if ((b > c) && (b < d))
202               :          0 :            max = d;
203  [ # # ][ # # ]:          0 :        else if ((b < c) && (b > d))
204               :          0 :            max = c;
205               :            :        else
206        [ # # ]:          0 :            if (c > d)
207               :          0 :                max = c;
208               :            :            else
209               :          0 :                max = d;
210               :            :
211               :          2 :    r[MAX_VALUE] = max;
```

Fig. 4.3 SC_THREAD find_max with the original coverage results for tests 1 and 2

by a + with blue color, and the F becomes − in color red (so not hit). Similarly, if another SW test evaluates the same condition with a false, the F becomes a + in blue (cf. Line 187 of Fig. 4.3 after execution of Test 1 (5 > 8 gave false) and Test 2 (5 > 3 gave true). This means that the expression has been tested by SW tests for both true and false cases. If the condition is not evaluated in any execution, it is marked with # (see, e.g., c>d of Line 194 of Fig. 4.3).

Demonstration of Consistency Analysis

When Test 1 is executed, the original IP finds the maximum value as 8, so this single test passes. In the following, we show concrete examples for categories C2 and C4.

C2 Example

Let us now assume that we mutate the IP in Line 191 in Fig. 4.4 by negating the complete if-condition. Then, running Test 1, this if-condition now results in false instead of true as before, and the execution jumps to the next *if*-statement (Line 194) to find the correct answer. Hence, the mutation causes the change of coverage visible at the statement in Line 194 in Fig. 4.4. Since the maximum of the inputs 5,1,8,4 is still 8 and will be also found by the current mutated IP, the test passes. Hence, we have fluctuating coverage, but passing SW tests, so this example falls in category

```
180          Branch      :   Line   :
181  [T    F][T    F]:       2 :     a = r[SRC_A] ;
182                   :       2 :     b = r[SRC_B] ;
183                   :       2 :     c = r[SRC_C] ;
184                   :       2 :     d = r[SRC_D] ;
185                   :         :
186          [ + - ]:       2 :     if (a > b)
187  [ + + ][ - + ]:       2 :         if ((a > c) && (a > d) )
188                   :       0 :             max = a;
189  [ + + ][ - - ]:       2 :         else if ((a > c) && (a < d))
190                   :       1 :             max = d;
191  [ + - ][ - + ]:       1 :         else if (!((a < c) && (a > d))) //MC2 - negate expression
192                   :       0 :             max = c;
193                   :         :         else
194          [ + - ]:       1 :             if (c > d)
195                   :       1 :                 max = c;
196                   :         :             else
197                   :       2 :                 max = d;
198                   :         :     else
199  [ # # ][ # # ]:       0 :         if ((b > c) && (b > d))
200                   :       0 :             max = b;
201  [ # # ][ # # ]:       0 :         else if ((b > c) && (b < d))
202                   :       0 :             max = d;
203  [ # # ][ # # ]:       0 :         else if ((b < c) && (b > d))
204                   :       0 :             max = c;
205                   :         :         else
206          [ # # ]:       0 :             if (c > d)
207                   :       0 :                     max = c;
208                   :         :             else
209                   :       0 :                 max = d;
210                   :         :
211                   :       2 :     r[MAX_VALUE] = max;
```

Fig. 4.4 Consistency ex.: coverage results for C2 example

C2. To solve the problem, the methodology guides the engineer to add a new test to the testsuite (Listing 4.2), e.g., Test 3 as depicted in Listing 4.3. With this new testsuite, the error is detected as the mutant calculates the wrong maximum value of 4 for the inputs 2,1,4,5. Therefore, the test fails (Line 7 in Listing 4.3), and so the considered mutation finally falls in category C1. Evaluating the same IP using classical mutation based qualification technique, an alive mutant will have to be chosen out of many. Hence, our method has reduced the search space.

C4 Example

For this example, we mutate the original *maxIP* block in Line 189 of Fig. 4.5 by replacing && with || operator. Test 2 with inputs 5,1,3,6 from Listing 4.2 results in a correct maximum value of 6, with stable line coverage. Hence, this example falls in category C4. Due to the presence of C3 mutants (which are not shown here), we know the propagation is problematic. Changing the coverage metric to branch coverage helps solving the problem. The branch coverage in Line 189 in Fig. 4.5 ([+ +][- +]) is now different from branch coverage in Line 189 in Fig. 4.3 ([+ +][+ -]). The C4 mutant therefore now becomes C2. So we have to add another test to the testsuite (Listing 4.2), e.g., Test 4 as depicted in Listing 4.4. With this new testsuite,

```
1   /* Test 3 */
2   ipBlock->SRC_A = 2;
3   ipBlock->SRC_B = 1;
4   ipBlock->SRC_C = 4;
5   ipBlock->SRC_D = 5;
6   if (ipBlock->MAX_VALUE == 5)
         success(3);
7   else fail(3);
```

Listing 4.3 Consistency ex.: SW test to kill mutant MC2

```
1   /* Test 4 */
2   ipBlock->SRC_A = 5;
3   ipBlock->SRC_B = 1;
4   ipBlock->SRC_C = 8;
5   ipBlock->SRC_D = 7;
6   if (ipBlock->MAX_VALUE == 8)
         success(4);
7   else fail(4);
```

Listing 4.4 Consistency ex.: SW test to kill mutant MC4

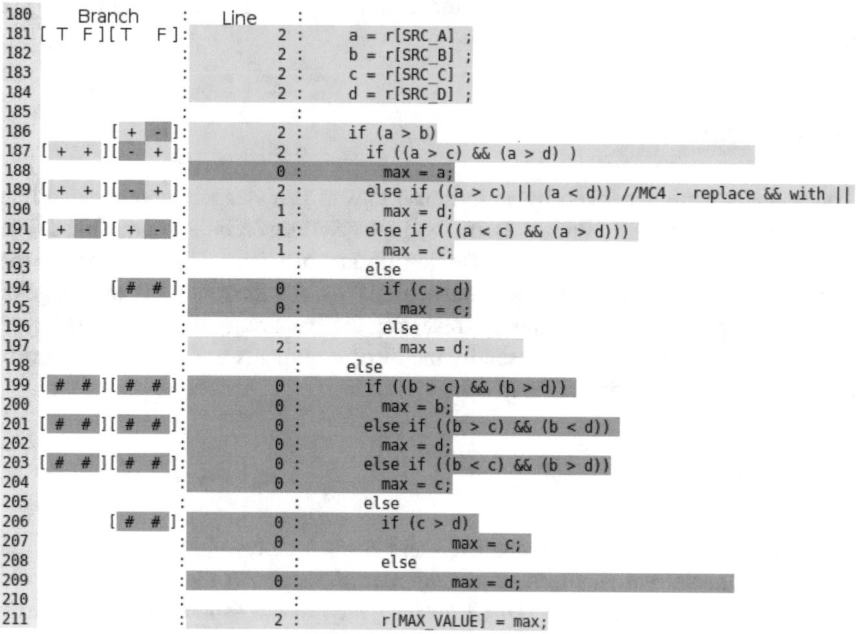

```
180      Branch        :   Line   :
181 [ T  F ][ T    F ]:       2 :    a = r[SRC_A] ;
182                    :       2 :    b = r[SRC_B] ;
183                    :       2 :    c = r[SRC_C] ;
184                    :       2 :    d = r[SRC_D] ;
185                    :         :
186          [ + - ]:         2 :    if (a > b)
187 [ + + ][ - + ]:       2 :      if ((a > c) && (a > d) )
188                    :       0 :        max = a;
189 [ + + ][ - + ]:       2 :      else if ((a > c) || (a < d))  //MC4 - replace && with ||
190                    :       1 :        max = d;
191 [ + - ][ + - ]:       1 :      else if (((a < c) && (a > d)))
192                    :       1 :        max = c;
193                    :         :      else
194          [ # # ]:         0 :        if (c > d)
195                    :       0 :          max = c;
196                    :         :        else
197                    :       2 :          max = d;
198                    :         :    else
199 [ # # ][ # # ]:       0 :      if ((b > c) && (b > d))
200                    :       0 :        max = b;
201 [ # # ][ # # ]:       0 :      else if ((b > c) && (b < d))
202                    :       0 :        max = d;
203 [ # # ][ # # ]:       0 :      else if ((b < c) && (b > d))
204                    :       0 :        max = c;
205                    :         :      else
206          [ # # ]:         0 :        if (c > d)
207                    :       0 :          max = c;
208                    :         :        else
209                    :       0 :          max = d;
210                    :         :
211                    :       2 :    r[MAX_VALUE] = max;
```

Fig. 4.5 Consistency ex.: coverage results for C4 example

the error is detected as the mutant calculates the wrong maximum value of 4 for the inputs 5,1,8,7. Therefore, the test fails (Line 7 in Listing 4.4), and so the considered mutation finally falls in category C1.

In the next section, the experimental results for our methodology for a real-world VP are given.

4.1.3   Experimental Results

This section presents the evaluation of our SW test qualification methodology in a software driven verification environment. We consider the LEON3-based VP SOCRocket [100] that has been modeled in SystemC TLM. We look at two IP integration scenarios, i.e., the integration of an *Interrupt Controller for Multiple Processors* (IRQMP) and the integration of a *General Purpose Timer* (GPTimer). In the following, we first describe how we automatically generate the mutants and the coverage models used in the case studies. Then, for each IP block, the basics are described before the demonstration of the methodology and qualification results are presented.

Mutant Generation

The mutants were generated by an in-house tool. It is a standalone command line tool that generates the mutants from the input SystemC/C++ source files. The underlying infrastructure is the LibTooling library of Clang. Clang generates the *Abstract Syntax Tree* (AST) for the input source files, and the tool takes advantage of that AST to compile new mutants by traversing different required entities. The type of mutants generated can be chosen by the input arguments to the tool. It supports all mutation operators described in Sect. 4.1.1.

Coverage Models

In addition to *Statement Coverage* (SC) and *Branch Coverage* (BC), we also use their strengthened variants termed as *Differential Statement Coverage (DSC)* and *Differential Branch Coverage (DBC)*, respectively, in the following. They signify the disturbance created by the mutant in terms of how many times the statement or branch was covered. The disturbance refers to the increase or decrease in coverage counters of statements and branches. It is calculated by taking the difference of coverage counters of the original model and mutant reported by the coverage tool LCOV. DSC and DBC are very useful for the elimination of mutants from categories C3 and C4 as these mutants often show stable statement and branch coverage.

IRQMP

Basics The first considered IP—IRQMP—processes incoming interrupts from different devices and processors based on priority. It supports 32 interrupt lines numbered from 0 to 31, where line 0 is reserved. Lines 1–15 are used for regular interrupts, whereas the remaining lines 16 to 31 for extended interrupts. The IRQMP model has a register file, I/O wires, and APB slave interface. The register file contains 32-bit processor-specific and configuration registers. When an interrupt is signaled, the corresponding bit is set in the register. This functionality is implemented using the SystemC thread *launch_irq* and callback functions, which are specified for register access (read/write).

The IRQMP interacts with connected processors by sending an interrupt request (*irq_req*) or receiving an acknowledgment (*irq_ack*). When an interrupt request is signaled for a processor, the IRQMP combines the `mask register` and the `pending register` with the `force register` to find the highest priority interrupt. The IRQMP also reads the `broadcast register` before forwarding the request to the processors. If the corresponding bit is set in `broadcast register`, the interrupt is broadcasted to all processors, i.e., written to the `force register` of all connected processors. In this scenario, the IRQMP expects acknowledgments from all processors. On the arrival of an interrupt request, if the corresponding bit is not set in `broadcast register`, it is simply set in the `pending register`. In this scenario, IRQMP expects an acknowledgment from any processor.

SW Test Qualification   The initial testsuite shipped with the IRQMP IP consists of 60 tests. This testsuite has 63% statement coverage of the IP. We add 45 tests to achieve high statement coverage (92%) as required by the methodology. The tool Typhon generates in total 244 mutants.

The results of applying our qualification methodology are shown in Table 4.4 where we report the first 13 iterations. The first row gives the index of the iteration. The second row states the operation done during those iterations, e.g., new tests are added to improve the testsuite, or the coverage metric is changed. The third row shows the metric used in the consistency analysis. The last row of Table 4.4 shows the consistency score of the SW testsuite calculated by using Eq. 4.1 for each iteration.

Handling C2 Mutants

The first iteration shows a low consistency score of 0.352. Due to the 14 mutants in category C2, it is clear that more tests need to be added to the suite. In the following, we describe one concrete mutant from category C2 to demonstrate its fix. An excerpt of the IRQMP model is shown in Listing 4.5. It shows the implementation of *incoming_irq* that handles the incoming interrupts from different devices. When such an interrupt arrives, the implementation checks the corresponding bit in `broadcast register` and handles the interrupt accordingly as described in the previous section.

```
1   void Irqmp::incoming_irq(const std::pair<uint32_t, bool> &irq, const
        sc_time &time) {
2   bool t = true;
3   if (!irq.second) {
4   // Return if the value turned to false. Interrupts will not be unset
5   // this way. So we can simply ignore a false value.
6     return;
7   }
8   for(int32_t line = 0 ; line<32; line++) {
9     if((1 << line) & irq.first) {
10    // Performance counter increase
11      m_irq_counter[line] = m_irq_counter[line] + 1;
12      v::debug << name() << "Interrupt line " << line << " triggered" <<
            v::endl;
13      if (!r[BROADCAST].bit_get(line)) {
14        r[IR_PENDING].bit_set(line, t);
15      }
16      if (r[BROADCAST].bit_get(line) && (line < 16)) { // Mutation: replace
            && with ||
17      // set force registers for broadcasted interrupts
18        for (int32_t cpu = 0; cpu < g_ncpu; cpu++) {
19          r[PROC_IR_FORCE(cpu)].bit_set(line, t);
20          forcereg[cpu] |= (t << line);
21        }
22      }
23    }
24  }
25  // Pending and force regs are set now.
26  // To call an explicit launch_irq signal is set here
27  e_signal.notify(2 * clock_cycle);
28  }
```

Listing 4.5 C2 IRQMP mutation example

When the mutation is performed (see comment in Line 16), the routine registers the interrupt in the `pending register` (Line 14) as well as in `force register` (Line 19). Thus, at least acknowledgment from at least one of the connected processors is expected. The SW tests, however, only check whether the interrupt was generated and handled. This is clearly a weakness of the existing tests. After the addition of a test, that checks for `broadcast register` and `pending register`, which leads to the SW test failure for this mutant. The mutant can thus according to our methodology be moved from C2 to C1. Moreover, the added test kills two more C2 mutants, resulting in 11 mutants in category C2 as can be seen in Iteration 2 of Table 4.4. Similarly, more C2 mutants are killed and moved to C1 by adding more tests from Iteration 3 to Iteration 10, where all C2 mutants are eliminated.

The importance of our methodology can be seen by looking at Table 4.4. Classical analysis has a search space of 131 mutants (#C2 + #C4), whereas our methodology has a limited search space of only 14 mutants. Hence, the guidance can save a lot of time by focusing the efforts on the right direction.

Table 4.4 IRQMP SW test qualification results

Iteration	1	2	3	4	5	6	7	8	9	10	11	12	13
Operation	Addition of tests										Change of metric		
Metric	SC										BC	DSC	DBC
Category C1	86	89	91	92	93	94	95	97	98	100	113	126	127
Category C2	14	11	9	8	7	6	5	3	2	0	43	98	99
Category C3	27	27	27	27	27	27	27	27	27	27	14	1	0
Category C4	117	117	117	117	117	117	117	117	117	117	74	19	18
Tests	105	106	107	108	109	110	111	112	113	114	114	114	114
Consistency score	0.352	0.365	0.373	0.377	0.381	0.385	0.389	0.398	0.402	0.410	0.463	0.516	0.520

SC Statement coverage, *BC* Branch coverage, *DSC* Differential SC, *DBC* Differential BC

Handling C3 and C4 Mutants

Now following the proposed methodology to eliminate the rest of mutants, we need to see both categories C3 and C4. The presence of only C3 mutants requires no action, but the presence of C4 mutants in conjunction indicates a problematic propagation. Hence, it is time to strengthen the coverage metrics used by the consistency analysis to eliminate C4 mutants in particular. We first strengthen the coverage metric to *BC* and apply consistency analysis. The results are shown in Table 4.4 as Iteration 11. As can be seen, the strengthening of coverage metric improves propagation, and thus, 43 mutants from C4 are moved into category C2, and additionally, 13 C3 mutations into category C1. In the next iterations, we deviated a bit from the methodology to demonstrate the effect of strengthening the coverage further. We changed the metric to *DSC* and then to *DBC*. The number of mutations in C4 and C3 category went down significantly as expected. The newly identified 99 C2 mutants now require more tests to be added.

Clearly, following the proposed methodology, weaknesses in the testsuite can be identified, and the test creator gets useful feedback on what to do next to improve the testsuite. The improvement is quantified by the consistency score, as a significant jump from 0.352 to 0.520 after 13 iterations can be observed.

GPTimer

Basics The second considered IP—GPTimer—implements down-counting timer(s) and generates an interrupt if zero is reached. The IP consists of 7 configurable timers that use ticks from the prescaler unit. The prescaler unit uses system clock as reference clock to decrement its value. The timer can also be configured to be used as a watchdog to prevent any malfunction. All the timers consist of a value register and a reload value register. When zero is reached or reset signal is initiated, the value register is loaded with the value in reload value register; otherwise, it is decremented by one in each cycle. The timers are not limited to only 2^{32} value but can also be executed for a longer duration by chaining them together. This way, the timers decrement when a zero is reached in the previous timer.

SW Test Qualification The testsuite shipped with the IP and is used as basis for our SW test qualification methodology consists of 14 tests initially.

For this IP, we report the results of the first 10 iterations as shown in Table 4.5. The terminologies used in Table 4.5 are the same as used in Table 4.4. The first iteration shows the distribution of mutants after *SC* analysis. Out of 308 mutants, 89 fall in category C1 initially and do not require any additional processing, whereas 219 mutants require additional work. Again, as expected, from iterations 2–6, the number of C2 mutants decreases as new tests are added to the suite. The total number of tests increases from 14 to 19. In Iteration 6, all mutants in C2 have been killed.

Table 4.5 GPTimer SW test qualification results

Iteration	1	2	3	4	5	6	7	8	9	10
Operation	Addition of tests						Change of metric			
Metric	SC						BC	DSC	DBC	PC
Category C1	89	91	92	93	94	95	131	139	140	182
Category C2	6	4	3	2	1	0	0	0	2	70
Category C3	112	112	112	112	112	112	76	68	67	25
Category C4	101	101	101	101	101	101	101	101	99	31
Tests	14	15	16	17	18	19	19	19	19	19
Consistency score	0.289	0.295	0.299	0.302	0.305	0.308	0.425	0.451	0.455	0.590

SC statement coverage, *BC* branch coverage, *PC* path coverage, *DSC* differential SC, *DBC*, differential BC

Therefore, following the methodology, we strengthen the coverage metric from *SC* to *BC* to try to eliminate category C4 and C3 mutants. The propagation is not observed for C4 mutants, but it can be observed for C3 mutants, as this moves 36 mutants from category C3 to C1. Changing the metric to *DBC* successfully detected two C2 mutants at the end of Iteration 9, where the consistency score has significantly increased from 0.289 to 0.455. There are still a big number of mutants in categories C4 and C3 (99 and 67, respectively). We decided to change the metric to path coverage to eliminate the mutants. As can be seen, the numbers reduced significantly from 99 to 31 (for C4) and from 67 to 25 (for C3), respectively.

In summary, the test creator can use our proposed methodology to strengthen the SW testsuite as he/she can now identify the weaknesses clearly. The strength of SW testsuite is shown in the last row of Table 4.5 as consistency score. It can be observed that the consistency score increased from 0.289 to 0.590 after 10 iterations.

4.2 Data Flow Testing for Digital Virtual Prototypes

Although a necessary step, statement coverage (and also stronger code coverage metrics) has some well-known limitations in their capability to detect bugs as well as to reflect the thoroughness of verification. As highlighted in the previous section, mutation analysis [26, 52] has been considered for decades in the SW testing community. The ideas have also been successfully transferred to hardware verification as well as to SystemC [16, 17, 47, 104]. Commercial mutation analysis tools with support for SystemC such as Certitude from Synopsys are also available.

Considering the successful adoption of mutation analysis, it is rather surprising that another effective testing technique known as *Data Flow Testing* (DFT) [82, 98] has not been yet considered for SystemC. DFT also holds the promise of better bug detection capability. The underlying idea of DFT is that the propagation of (wrong) data is a necessity to reveal bugs: If a line of code produces a wrong value, the execution after that point must include another line of code that uses this erroneous value; otherwise, there will be no observable failures. Based on this information, researchers have proposed several data flow adequacy criteria. These criteria require the testsuite to sufficiently exercise the identified *definition–use pairs*, i.e., pairs of definition (a statement where a value is produced) and use (a statement where this value is used). Recent research in DFT focused on automated test generation for data flow adequacy [111, 117] and on extending DFT to object-oriented programs [4, 27].

In this section, we present the first DFT approach for SystemC-based VPs, which does not come without challenges. A SystemC DUV is essentially a concurrent program with non-preemptive thread scheduling, shared memory communication, and event-based synchronization. This unique combination requires rethinking of the known DFT techniques. To this end, our contribution is twofold: First, we develop a set of SystemC specific coverage criteria for DFT, which takes the non-preemptive context switches and synchronization primitives of SystemC into consideration. Second, we explain how to automatically compute the data flow coverage result for a DUV using a combination of static and dynamic analyses. The coverage result provides clear suggestions for the testing engineer to add new test-cases in order to improve the coverage result. Our experimental results on real-world VPs demonstrate the applicability and efficacy of our analysis approach and the SystemC specific coverage criteria to improve the testsuite quality.

4.2.1 SystemC Running Example

We present here an example SystemC program (Listing 4.6) that will be used to showcase the main ideas of our approach throughout this section. The SystemC constructs and semantics necessary to understand the example will be explained as needed. The example consists of two modules *producer* and *consumer* that communicate through a FIFO. Their behavior is implemented in thread functions (producer: `prod_thread()` Line 37—consumer: `recv()` Line 53, `filter()` Line 68, `send()` Line 79) registered in the simulation kernel. The behavior of the consumer depends on the input provided by the producer. The FIFO provides a *write* and a *read* function that adds or removes an element, respectively. The write function is used from the producer thread in Line 41, and the read function from the consumer thread in Line 57. Both functions can potentially suspend the threads execution in case the FIFO is empty on read attempt (Line 16) or full at write attempt (Line 7). The thread becomes runnable again when the awaited event is notified in Line 11 or Line 21, respectively. The consumer module itself consists of three threads. The thread `recv()` is responsible to retrieve the next produced element x (Line 58) from the FIFO and transfers it to the send thread for processing. The transfer can happen in two ways: (1) Through the `filter()` thread that applies post-processing and checking (Line 68), or (2) directly without delay to the `send()` thread for high priority items (Line 79). However, the send thread only accepts one fast transfer

```
1   struct fifo : public sc_channel {                 50      SC_THREAD(recv); SC_THREAD(send);
2     fifo(sc_module_name name)                                   SC_THREAD(filter);
3       : sc_channel(name), num_elements(0), first(0) {}  51    }
4                                                       52
5     void write(char c) {                              53    void recv() {
6       if (num_elements == max)                        54      wait(0, SC_NS); // ensure send and filter are
7         wait(read_event);                                             run first
8                                                       55      char c;
9       data[(first + num_elements) % max] = c;         56      while (true) {
10      ++num_elements;                                 57        in−>read(c); // in is bound to fifo instance
11      write_event.notify();                           58        x = c;
12    }                                                 59        filter_event.notify(1, SC_NS);
13                                                      60        if (x < 10) {
14    void read(char &c){                               61          // high priortiy data handled immediately
15      if (num_elements == 0)                          62          send_fast_event.notify();
16        wait(write_event);                            63        }
17                                                      64        wait(recv_event);
18      c = data[first];                                65      }
19      −−num_elements;                                 66    }
20      first = (first + 1) % max;                      67
21      read_event.notify();                            68    void filter() {
22    }                                                 69      while (true) {
23                                                      70        wait(filter_event);
24  private:                                            71        if (x < 0)
25    enum e { max = 10 };                              72          x = 0;
26    char data[max];                                   73        if (x > 126)
27    int num_elements, first;                          74          x = 126;
28    sc_event write_event, read_event;                 75        send_regular_event.notify(1, SC_NS);
29  };                                                  76      }
30                                                      77    }
31                                                      78
32  SC_MODULE(producer) {                               79    void send() {
33    SC_CTOR(producer) {                               80      bool fast_mode = true;
34      SC_THREAD(prod_thread);                         81      while (true) {
35    }                                                 82        if (fast_mode) {
36                                                      83          wait(send_regular_event | send_fast_event);
37    void prod_thread() {                              84          fast_mode = false;
38      wait(0); // start together with the consumer    85        } else {
39      const char ∗str = "SystemC Example"; // input   86          wait(send_regular_event);
                       to the design, can influence consumer  87          fast_mode = true;
                       behavior                         88        }
40      while (∗str)                                    89        assert (x >= 0);
41        out−>write(∗str++); // out is bound to fifo   90        cout << x << endl;
                       instance                         91        recv_event.notify();
42    }                                                 92      }
43                                                      93    }
44    sc_port<fifo> out;                                94
45  };                                                  95    sc_port<fifo> in;
46                                                      96  private:
47                                                      97    int x;
48  SC_MODULE(consumer) {                               98    sc_event recv_event, filter_event,
49    SC_CTOR(consumer) {                                            send_regular_event, send_fast_event;
                                                        99  };
```

Listing 4.6 SystemC example

at a time (Line 80, controlled by the `fast_mode` variable). Therefore, the filtering is always unconditionally initiated (Line 59) as fallback in case the send thread currently does not support fast processing. Finally, the `send()` thread will notify the `recv()` thread (Line 91) to transfer the next element.

4.2.2   Def–Use Association and Data Flow Testing

A def–use association is an ordered triple (x, d, u) such that d is a statement where variable x is defined and u is a statement where x is used. Furthermore, there is a path in the program from d to u without re-definition of x. For example, consider Listing 4.6: variable `fast_mode` is defined in Line 80 and used in Line 82, so it is a def–use association. A def–use association (x, d, u) is exercised by a test-case t, iff execution of t goes through definition d and then uses u without re-definition of variable x in between.

Data flow testing tries to maximize the exercised def–use associations. Essentially, it works by refining the testsuite by adding test-cases until the coverage criteria are met or testing resources are exhausted. This requires detecting def–use associations and measures the data flow coverage of the current testsuite. How to do this for SystemC, and thereby taking SystemC specifics into account, is shown in the following. Please note that we will use the term data flow association as a generalization of a def–use association to avoid confusion. The reason is that we define a SystemC specific wait–notify association, which is also a data flow association.

4.2.3   Data Flow Testing for SystemC

Overview

An overview of our data flow testing approach for SystemC is shown in Fig. 4.6. Essentially, our approach combines a static and dynamic analysis to fully automatically compute a SystemC specific data flow coverage result.

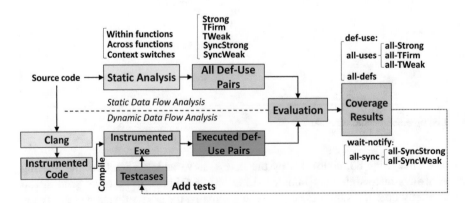

Fig. 4.6 An overview of our data flow testing approach for SystemC

The *static analysis* (upper half of Fig. 4.6) identifies the set of all data flow associations. Our static analysis computes an over-approximation of all def–use associations and thus also contains *infeasible* associations, i.e., associations that cannot be exercised no matter which input is applied. A precise computation of all feasible associations requires heavy use of formal verification techniques and therefore is not practical due to scalability issues. To guide test-case selection, associations are classified into different disjoint groups based on the likeliness of being infeasible. Please note that the static analysis needs to be only run once at the beginning on the source code.

The *dynamic analysis* (lower half of Fig. 4.6) detects which data flow associations have been exercised by the current testsuite. It works by instrumenting the SystemC source file to log relevant run-time information. The instrumented source file is then compiled with a standard C++ compiler and executed for every test-case. The resulting logs are analyzed and combined to obtain the set of exercised data flow associations.

In the next step, both static and dynamic analysis results are evaluated and combined to obtain a coverage result. Essentially, the result shows which data flow associations have been exercised by at least one test-case and which have been completely missed. An association can be missed due to two reasons: (1) The testsuite is insufficient to cover it. In this case, a new test-case needs to be added. (2) The association is infeasible, i.e., there is no possible input that will cover it. In this case, it can be ignored.

Our classification system, which ranks associations according to their likeliness of being infeasible, allows the testing engineer to focus his efforts on promising test-cases to efficiently improve the coverage result. Please note that we do not yet consider automated test generation to exercise a specific data flow association in this work, as it will exceed the scope of this book. Automated test generation is left for future work.

In the following, we describe our classification system and the coverage result in more detail and demonstrate both using the running example.

Classification of Data Flow Associations

Our classification system attempts to preserve scalability of the data flow testing approach and at the same time provide meaningful results and suggestions to guide the test-case generation. We define *five SystemC specific classifications*: Strong, *TFirm, TWeak, SyncStrong*, and *SyncWeak*.

The first three (details see below) extend the classical notion of data flow testing that reason about variable definition and use. Therefore, these fall into the def–use association category. These classifications especially deal with the non-preemptive threads of the SystemC simulation (and hence, the T in TWeak and TFirm).

The last two classifications are used to classify event-based synchronization of SystemC by means of the wait–notify function. This can also be considered as a data flow relation. The wait can be considered a definition that suspends the active thread, while the notify is considered a use. However, to avoid confusion, we introduce a new data flow association called *wait–notify association* for these synchronization related flows.

Def–Use Associations Our static analysis reports the following def–use associations (x, d, u): There is a static path from d to u in the program without re-definition of x in between. Please note that there is a static path from every context switch statement from one thread to the start of a transition of every other thread. A transition starts at the beginning of a thread and right after a context switch.

Based on this general observation, we define three classifications for def–use associations (x, d, u). In this context, we define a *du-path* as a static path between d and u without re-definition of x:

- *Strong* : (a) Every *du-path* is without context switch, or (b) x is a thread local variable, or (c) d is the only definition of x, i.e., x is a constant.
- *TFirm* : At least one *du-path* is without context switch and at least one *du-path* with context switch.
- *TWeak* : Every *du-path* contains a context switch.

Since a local thread variable cannot be redefined on context switch, the def–use associations are considered Strong. This refinement of def–use associations provides clear guidelines for test-case selection. In general, one should focus on Strong and TFirm associations first. In both cases, there exists at least one (static) path without context switch in between the definition and use.

Wait–Notify Associations Similarly to def–use associations, we define a wait–notify association as an ordered triple (e, w, n) where w contains a wait of event e and n contains a notify of event e, and there exists a path in the program such that w is notified from n.

This definition does not require a notify to happen after the wait during execution. This is due to timed notification, where the notify statement only schedules the notification to happen at a later point in time. Such a timed notification is much more difficult to handle precisely using static information. Furthermore, the notification can be canceled, e.g., by issuing a new notification of event e. Immediate notifications on the other hand more directly resemble the classical data flow relation as the notification happens when the notify statement is executed. Events are inherently not thread local, and there is always a context switch between a wait and a notify, which makes static analysis more difficult.

Our static analysis detects the following wait–notify associations (e, w, n): (1) A timed notify n for event e is executed, and the scheduled notification is not canceled until a context switch is executed. And w is a wait for event e. (2) w is a wait for event e in one thread, and n is an immediate notify of event e on another thread.

This approximation can be further refined by applying an additional analysis that computes transitions between threads more precisely. However, this approximation is sufficient when dealing with SystemC synchronization primitives. One of the reasons is that often only very few wait–notify statements operate on a single event. Therefore, it is reasonable to assume that most of them are feasible.

This observation motivates to introduce the following two classifications for wait–notify associations (e, w, n):

- *SyncStrong*: Wait w should have a one-to-one relationship with a notify n for an event e.
- *SyncWeak*: Otherwise, i.e., multiple wait or notify statements are available for event e.

Coverage Result

Every classification defines a disjoint set of data flow associations. Therefore, we define a coverage criterion for each classification. For instance, the all-Strong criterion is satisfied, iff all data flow associations classified as Strong have been exercised. Criteria for the remaining classifications—i.e., all-TFirm, all-TWeak, all-SyncStrong, and all-SyncWeak—can be defined analogously.

Based on these criteria, we can define specific def–use and wait–notify criteria:

- The all-uses criterion requires that all-Strong, all-TFirm, and all-TWeak are satisfied.
- The all-defs criterion requires that for every definition D in the program at least one def–use association (x, d, u) with $D = d$ is exercised.
- The all-sync criterion requires that all-SyncStrong and all-SyncWeak are satisfied.

Finally, the all data flow criterion is satisfied iff all-uses and all-sync criteria are satisfied.

While in general satisfying all data flow criterion is not practical due to imprecisions in the static analysis, which can result in infeasible data flow associations, it is possible that some of the (sub-)criteria can be fully satisfied—or at least up to a high degree, i.e., 95% of the associations have been exercised. In particular, we expect that all-Strong, all-TFirm, and all-SyncStrong to be the primarily focused criteria.

Both Strong and TFirm associations contain a (static) path without re-definition and without context switch between the definition and use. Therefore, we expect that a test-case will exercise them. Otherwise, the definition is dead code or all relevant paths (without re-definition of variable x) between definition and use are infeasible.

Similarly, if the SyncStrong criterion is not satisfied, it means that some notification has never reached a wait, or some wait has not been notified. This implies that some wait–notify statement is essentially unused.

Table 4.6 Data flow associations for the SystemC example in Listing 4.6 sorted by classification

Strong				
(c,5,9)	X	X	X	X
(c,57,58)	X	X	X	X
(fast_mode,80,82)	X	X	X	X
(fast_mode,84,82)	X	X	X	X
(fast_mode,87,82)	X	X	X	X
(max,25,6)	X	X	X	X
(max,25,9)	X	X	X	X
(max,25,20)	X	X	X	X
(num_elements,3,6)	X	X	X	X
(num_elements,3,15)	X	X	X	X
(num_elements,10,6)	X	X	X	X
(num_elements,19,15)	X	X	X	X
(str,39,40)	X	X	X	X
(str,39,41)	X	X	X	X
(str,41,40)	X	X	X	X
(str,41,41)	X	X	X	X
(x,72,73)		X		
(x,58,60)	X	X	X	X

TFirm				
(data,3,18)	–	–	–	–
(first,3,9)	X	X	X	X
(first,3,18)	X	X	X	X
(first,3,20)	X	X	X	X
(first,20,18)	X	X	X	X
(first,20,20)	X	X	X	X
(num_elements,3,9)	X	X	X	X
(num_elements,3,10)	X	X	X	X
(num_elements,3,19)	–	–	–	–
(num_elements,10,9)	X	X	X	X
(num_elements,10,10)	X	X	X	X
(num_elements,19,19)	X	X	X	X

SyncWeak				
(send_regular_event,83,75)	X	X	X	X
(send_regular_event,86,75)	X	X	X	X

TWeak				
(data,9,18)	X	X	X	X
(first,20,9)	X			
(num_elements,10,15)	X			
(num_elements,10,19)	X	X	X	X
(num_elements,19,6)	–	–	–	–
(num_elements,19,9)	X			
(num_elements,19,10)	X			
(x,72,71)	–	–	–	–
(x,72,89)		X		
(x,72,90)		X		
(x,74,71)	–	–	–	–
(x,74,73)	–	–	–	–
(x,74,89)		X		
(x,74,90)		X		
(x,58,71)	X	X	X	X
(x,58,73)	X	X	X	X
(x,58,89)	X	X	X	X
(x,58,90)	X	X	X	X

SyncStrong				
(filter_event,70,59)	X	X	X	X
(read_event,7,21)	X			
(recv_event,64,91)	X	X	X	X
(send_fast_event,83,62)				X
(write_event,16,11)	X	X	X	X

Illustration

We have executed the SystemC example in Listing 4.6 with four different inputs to gradually increase the data flow coverage. In particular, we used the following inputs for the str variable in Line 39: (1) "SystemC Example," (2) "Abc\x7f," (3) "a\tb\xff," and (4) "Test\x80." Table 4.6 shows the results, i.e., the statically classified data flow associations, and by which test-case they have been exercised (marked by X). Infeasible data flow associations are marked with "-" for all test-cases. The information is grouped into four main columns and is read from top to bottom and then from left to right. In this order, the Strong, TFirm, TWeak, SyncStrong, and SyncWeak associations are listed.

For example, consider the Strong def–use association ($num\_elements$, 10, 6) shown in the first column and exercised by all test-cases. This one is Strong because all paths between Lines 10 and 6 are without re-definition of num_elements and are free of context switches. In fact, there is only one possible path from Line 10 to Line 6—first the write function is exited, then the while loop is not finished, and so the write function is called again.

For the def–use association ($num\_elements$, 10, 9), there are two paths between Line 10 and Line 9, due to the branch in Line 6. One involves a context switch and the other path does not. Therefore, this association is TFirm.

The def–use association ($first$, 20, 9) is TWeak because the only way to reach the use starting from the definition is through a context switch. This association is only exercised by the first test-case.

The associations involving the max variable are classified Strong because it is a constant, so it cannot be redefined. The association (c, 5, 9) is also Strong, even though there are two paths from 5 to 9, and one involves a context switch because c is thread local—so it cannot be redefined due to the context switch. Most wait–notify associations are SyncStrong because there is only a single wait and notify for the event. The ports in and out are bound in the elaboration phase, which is not shown in the example. Therefore, no def–use association is reported for them in Table 4.6.

The first test-case already achieves a reasonable data flow coverage, in particular for the FIFO component, which works independent of the actual input. Since the all-Strong and all-TFirm coverage criteria are already satisfied, the next step is to consider the TWeak associations.

The second and third inputs use special characters that are processed separately in the filter thread in Lines 71–74. With the first three inputs, all feasible def–use associations are covered. Therefore, the maximal def–use coverage has been achieved for this example. Please note that full branch and statement coverage has also been achieved with this test set.

However, the coverage with regard to wait–notify associations can still be increased. In particular, the association ($send\_fast\_event$, 83, 62) has not been exercised. The reason is that with the current testsuite all special characters were passed through the filter thread since the send thread has not been in $fast\_mode$, i.e., waiting in Line 86. Therefore, we have added a fourth test-case to exercise its association. This test-case was able to detect a bug in the design, where an invalid character is passed to the send thread.

This example demonstrates that the standard def–use coverage criteria alone are not sufficient for extensive testing of SystemC designs. Our proposed SystemC specific wait–notify coverage criteria is important for a high quality testsuite.

4.2.4   Implementation Details

This section describes the static and dynamic analysis of our data flow testing framework in more details. As aforementioned, the static analysis computes an over-approximation of data flow associations classified into disjoint groups. The dynamic analysis then detects which data flow associations really have been exercised by the testsuite.

Static Analysis

The static analysis is implemented using the LibTooling library for Clang compiler. Clang generates an *Abstract Syntax Tree* (AST) of the SystemC source code. The AST is parsed to extract the required information to perform static analysis. Then it applies three subsequent analysis steps: (1) Local analysis within every function, (2) Information propagation across function calls, and (3) Consideration of the effects of context switches.

Dynamic Analysis

The dynamic analysis works in two steps: (1) Instrument the SystemC source code using the Clang compiler framework to log data flow relevant information. Then execute all test-cases on the instrumented executable to generate the log. (2) Analyze the log line by line to build the exercised data flow associations. Both steps are described in the following.

Source Code Instrumentation

In the first step, the SystemC source code is instrumented to log data flow relevant information. Therefore, it is analyzed statement by statement to detect (1) definitions and uses of variables and (2) wait and notify of events. For every such detected information, a print instruction, which writes to a log, is placed before the statement. Please note that the order of the print instructions is important in case multiple information are available, e.g., $i++;$, in general the uses are placed before the definitions. For the while loop (and similarly the for loop), print instructions for the loop condition are replicated at the end of the loop since the condition is re-evaluated in every loop step. Therefore, it is not enough to place them only before the while loop.

Def–Use Associations Construction

The def–use associations can be identified in a straightforward way. We keep a mapping of active definitions. It relates each variable to its last definition, which is updated on re-definition. Whenever a use for a variable is found in the log, the corresponding definition is retrieved from the mapping.

Wait–Notify Associations Construction

Detecting wait–notify associations requires additional work due to timed notifications. This decouples the notify statement from the actual event triggering. Therefore, we modified the SystemC kernel to write a log entry whenever an event is triggered.

For example, consider the statement sequence *e.notify(1, SC_NS); wait(e);*, where an event *e* is scheduled for notification, and then the thread is suspended to wait for the notification *e*. The execution log contains a notify entry for event *e* followed by a wait entry for *e*. The actual trigger from the SystemC kernel appears later. In between can be other log entries. Other threads can also wait for event *e*. However, a new notification for *e* will cancel the one before.

Based on this information, we keep a mapping from an event to a set of active wait statements and a mapping to the last scheduled notification. Once the kernel trigger is parsed for an event *e*, retrieve the last notification *n* and a set of active waits *S*. Then add a wait–notify association (e, w, n) for every $w \in S$. Immediate notifications are handled in the same manner; the trigger simply appears directly after the notification scheduling in the log.

4.2.5   Experimental Results

In this section, we present a case study to demonstrate our DFT approach for SystemC. We consider the *Interrupt Controller for Multiple Processors* (IRQMP) component from the LEON3-based VP SOCRocket [100] discussed earlier in Sect. 4.1.3. The initial testsuite shipped with the IRQMP component consists of 60 tests achieving 63% statement coverage, 74% branch coverage, and 63% data flow coverage: a total number of 571 data flow associations have been computed by our static analysis, and 359 of them have been found to be exercised by the testsuite using our dynamic analysis. Initially, 62% Strong, 71% TWeak, and 58% SyncWeak data flow associations are exercised. There is no TFirm association as each instruction has to pass *wait* statement; hence, a context switch is inevitable. The testsuite does not fulfill the all-uses, all-defs, and all-sync criteria. Therefore, the all data flow criterion is not satisfied. One interesting point is that there are no SyncStrong wait–notify pairs; instead, there are 5 SyncWeak wait–notify pairs w.r.t. the only *sc_event e_signal*.

To increase the def–use coverage, the uncovered 38% Strong associations are satisfied first by adding additional tests manually. They are followed by 29% TWeak and 42% SyncWeak associations. In total, 54 additional tests are added to the initial testsuite. These new tests increased the overall data flow coverage to 88%, with 96% Strong, 91% TWeak, and 69% SyncWeak data flow associations exercised. After this point, we found it to be very hard to improve these numbers further. This demonstrates the need for future research on automated test generation techniques for SystemC that are capable to derive hard-to-find tests and to prove infeasibility of associations. With the now enhanced testsuite, we also achieve 92% statement coverage and 89% branch coverage. Despite the very high values of these code coverage metrics, they provide no insight into potential synchronization issues. In contrast, our developed data flow coverage with regard to wait–notify association shows that in many cases there was no *wait* statement waiting for notify because some other function had already fulfilled the *wait* condition. This can lead to potential problems in the integration of the component in a larger system/subsystem, e.g., an incoming interrupt that needs to be handled quickly can be delayed.

4.3   Data Flow Testing for SystemC AMS Virtual Prototypes

This section goes beyond purely digital VPs and extends DFT for AMS VPs, in particular SystemC-AMS *Timed Data Flow* (TDF) models. We propose the first DFT approach for SystemC-AMS TDF models based on two major contributions: First, we develop a set of SystemC-AMS TDF models specific coverage criteria for DFT. This requires considering the SystemC-AMS semantics of TDF signals, TDF cluster modeling, TDF ports, and dynamic TDF. Second, we explain how to automatically compute the data flow coverage result for given TDF models using a combination of static and dynamic analysis techniques. The coverage results guide the verification engineer to add new test-cases to improve the coverage.

4.3.1   SystemC AMS Motivating Example

We present here a simplified example SystemC-AMS program (Listing 4.7) extracted from an IOT device, *sensor_system* (Fig. 4.7). This example will be used to showcase the main ideas of our approach throughout this section. The *sensor_system* is a typical AMS system that includes a mixture of analog and digital components: a *Temperature Sensor* (TS), a *Humidity Sensor* (HS), an analog signal delay (Z^{-1}), a 4×1 *Analog mux* (AMUX), a gain element (G), a 9-bit *Analog-to-Digital Converter* (ADC), a small digital control module, and two *Light-Emitting Diodes* (LEDs). The *sensor_system* works as follows: TS/HS senses the temperature/humidity (time continuous analog signals), and if the sensed value crosses a certain threshold, the sensor generates an interrupt (digital signal) to the control module. The control module sets the AMUX select line corresponding to the

```
1    void TS::processing()                        41    void ctrl::processing()
2    {                                            42    {
3      double sig_in = ip_signal_in; // volts    43      if(ip_intr0)
4      double tmpr = sig_in*1000; //millivolts   44        if((ip_DIN/10) < 60) {
5      double out_tmpr = 0;                       45          op_clear = 1;
6      bool intr_ = false;                        46          m_mux_s = 0;
7      if (!ip_hold){                             47          op_hold = 0;
8        if (ip_clear) intr_ = 0;                 48        } else if (m_mux_s == 1 &&
9        else if ((tmpr > 30) && (tmpr < 1500 )){          (ip_DIN/10)>60){
10         out_tmpr = tmpr;                       49          op_T_LED = 1;
11         intr_ = true;                          50          op_clear = 1;
12       }                                        51          op_hold = 0;
13       op_intr.write(intr_);                    52          m_mux_s = 0;
14       op_signal_out = out_tmpr;                53        } else if (m_mux_s == 0 && (ip_DIN/10)>50){
15     }                                          54          m_mux_s = 1;
16   }                                            55          op_hold = 1;
17                                                56        } else {
18   void HS::processing()                        57          op_hold = 0;
19   {                                            58          op_clear = 1;
20     double temp = ip_signal_in*1000; // mV     59          m_mux_s = 0;
21     double Tdepend = (B1*42 + B2)*temp +       60        }
                 (B3*42+B4);                      61      else if (ip_intr1 && m_mux_s == 2){
22     double C = 153e-12; // capacitance         62        if(ip_DIN > 45) op_H_LED = 1;
23     double BC = 150e-12; // bulk capacitance at 63      m_mux_s = 0;
                 30%RH                            64      } else if (ip_intr1)
24     double sensitivity = 0.25e-12;             65        m_mux_s = 2;
25     bool intr_ = false;                        66      op_mux_s = m_mux_s;
26     double newRH = 30 + ((C − BC)/sensitivity) + 67    if(ip_intr0==0) op_clear = 0;
                 Tdepend;                         68    }
27     if (newRH > 30) intr_ = true;              69
28     op_intr.write( intr_);                     70    void sense_top::architecture() // netlist
29     op_signal_out = newRH;                     71    {
30   }                                            72      // .....
31                                                73      i_delay_tdf1 −> tdf_i.bind(op_signal_out);
32   void AM::processing()                        74      i_delay_tdf1 −> tdf_o.bind(op_delay_out);
33   {                                            75
34     double tmp_out = 0;                        76      i_gain_tdf1 −> tdf_i.bind(op_mux_out);
35     if (ip_select == 0) tmp_out = ip_port_0;   77      i_gain_tdf1 −> tdf_o.bind(op_gain_out);
36     else if (ip_select == 1) tmp_out = ip_port_1; 78
37     else if (ip_select == 2) tmp_out = ip_port_2; 79    i_adc1 −> adc_i.bind(op_gain_out);
38     op_mux_out = tmp_out;                      80      i_adc1 −> adc_o.bind(op_adc_out);
39   }                                            81      //...
40                                                82    }
```

Listing 4.7 Sensor system - SystemC-AMS TDF model example (B1 = 0.0014/°C, B2 = 0.1325 RH/°C, B3 = -0.0317, B4 = -3.0876 RH)[28]

sensor (Line 66) and reads the temperature/humidity value. It compares the value with a preset threshold (60 °C for temperature, 45RH for humidity) and switches the corresponding LED (T_LED (Line 49)/H_LED (Line 62)) on, indicating too hot or too humid. Please note that if the temperature crosses 50 °C, the controller halts the sensor output (Line 55) and reads the delayed value to ensure it reads the value correctly in the first place. The whole system is implemented in SystemC-AMS TDF MoC, where the necessary parts of the design are shown in Listing 4.7. Please note that the variables with prefix "ip_"/"op_" and "m_" refer to input–output ports, and member variables of the TDF model, respectively.

The TS is implemented in function *TS::processing()* (Lines 1–16). It gives an output signal only when the input is between a certain threshold [between 30 mV

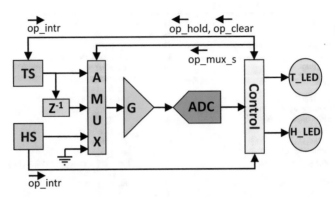

Fig. 4.7 Sensor system—T: Temperature, H: Humidity, XS: X Sensor (X=T,H), Z^{-1} = delay, AMUX: Analog mux, ADC: Analog-to-Digital Converter, X_LED: Light-emitting Diode, G: Gain

and 1500 mV (Line 9)]. HS behavior is implemented in *HS::processing()* (Lines 18–30). Interrupts are additionally added to both the sensors. AMUX is implemented in *AM::processing()* (Lines 32–39), and control in *ctrl::processing()* (Lines 41–68). The control model translates incoming digital signal from ADC into a temperature reading by dividing the input (ip_DIN) by 10 (the scale factor). For example, a signal of 200 mV translates into 20 °C. Gain (G) and analog delay (Z^{-1}) (SystemC-AMS library) are instantiated as shown in Listing 4.7 at Lines 76 and 73, respectively. ADC is instantiated at Line 79. In order to simplify the showcase of example, the ADC is assumed to output the same signal in digital form, and it gets on its analog input. A required part of netlist (binding information) of the TDF cluster is shown in *sense_top::architecture()* (Lines 70–82). In the following sections, data flow associations and testing w.r.t. SystemC-AMS TDF models will be shown while considering the TDF semantics.

4.3.2 Data Flow Testing for SystemC-AMS TDF Models

In this section, we start with the overview of our approach. Afterward, we provide the ingredients such as classification of data flow associations w.r.t. SystemC-AMS TDF models, and coverage criteria. At the end, the approach is illustrated to show its effectiveness.

Approach Overview

Our data flow testing approach for SystemC-AMS TDF models is shown in Fig. 4.8. It comprises three stages: (1) static analysis, (2) dynamic analysis, and (3) coverage

Fig. 4.8 Proposed data flow
testing methodology
overview

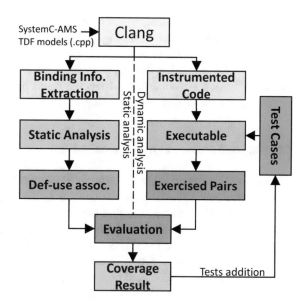

analysis. All three stages work together to fully automatically compute SystemC-
AMS TDF models specific data flow coverage results.

The static analysis identifies the set of all data flow associations computed in the
TDF clusters using TDF models binding information [13]. This step is executed only
once at the beginning of the analysis. Due to static nature, the analysis computes an
over-approximation of the associations that may also contain infeasible associations,
i.e., dead code associations. On circuit level, dead code associations can be mapped
to component isolation because of open circuit, wrong transistor configuration, or a
very high/low passive component value, etc. To guide test-case selection (with test
input signal), associations are classified into different disjoint groups based on the
likeliness of being infeasible.

The dynamic analysis (Fig. 4.8—right side) identifies the exercised data flow
associations w.r.t. the testsuite. The SystemC-AMS source file with TDF models is instrumented so that relevant run-time information can be captured. The
instrumented file is afterward compiled and executed against every test-case from
testsuite. For each test-case, different data (signal) flow information is recorded.
The resulting logs are analyzed and combined to obtain the set of exercised data
flow associations.

At the end, the static and dynamic analysis results are evaluated and combined to
obtain a coverage result. Essentially, the result shows which data flow associations
have been exercised by at least one test-case and which have been completely
missed. An association can be missed due to: (1) The testsuite is insufficient to
cover it. In this case, a new test-case (test input signal with different parameters)
needs to be added. (2) The association is infeasible. In this case, it can be either
inspected for correct binding or ignored.

Our classification system, which ranks associations according to their likeliness of being infeasible, allows the testing engineer to focus his efforts on promising testcases (test input signals) to efficiently improve the coverage result. In this chapter, automated test generation has not been considered. In the following, we describe our classification system and the coverage result in more detail and demonstrate both using the running example (Listing 4.7).

Classification of Data Flow Associations

We define four classifications specific to TDF models of SystemC-AMS: Strong, Firm, PFirm, and PWeak. They are defined in two categories, (1) within a TDF model, (2) in TDF cluster. The first category extends the classical notion of data flow testing that reason about variable definition and use within one function. First two classifications come under this category. In SystemC-AMS, a variable defined in one TDF model might flow to another TDF model for use. This happens when the defined variable is an output port, and the used variable is an input port. In this case, the classical notion does not apply, and new approach has to be devised. In this scenario, it is also possible that the defined variable (port) is redefined (signal amplification, signal delay, etc.) outside the TDF model, before being used. Hence, the two classifications, PFirm, PWeak, are used to classify the definition and use w.r.t. the SystemC-AMS TDF model ports (hence, P in PFirm and PWeak).

Def–Use Association We define a def–use association as an ordered tuple (v,d,dm,u,um). For a variable v, there exists a static path between definition d present in the TDF model dm (*defining model*) and use u present in TDF model um (*using model*) without a re-definition of v in between. We define a *du-path* as a static path between d and u without a re-definition of v, when both d and u can be in same or different TDF models. Four classifications for def–use associations are proposed:

1. *Strong*—(a) Variable v is an output port of a TDF model dm, and there exists a *du-path* between dm and um. b) Variable v is local to the TDF model, i.e., $dm ==um$, and every static path between *def* and *use* is a *du-path*.
2. *Firm*—Variable v is local to the TDF model, and at least one static path between d and u is not a *du-path*.
3. *PFirm*—Variable v is an output port, and there exists at least one static path that is not a *du-path*.
4. *PWeak*—Variable v is an output port, and there exists no *du-path*.

Because local variable does not flow out of the TDF model, it is classified into first two associations: strong or firm. When ports of TDF models are considered, the data (signal) flowing toward the connecting TDF model might be a *du-path* (direct connection) to the *um*. It is also possible that the data (signal) is redefined. The re-definition could occur in the following two cases (limited to SystemC-AMS library, and only to *Single Input Single Output* (SISO) components): (a) There exists a *delay* element in the *du-path*. The delay element delays the incoming data (signal) by the preset number of samples (seconds) and outputs an earlier value instead of the current value. Because of this reason, we consider it as a re-definition. (b) There exists a *gain or buffer* element that amplifies the incoming signal or regenerates it. More elements can also be added. Based on this general idea, three data flow associations are defined: (a) If the output port directly connects to another TDF model, the association is *Strong*. (b) If the output port connects to another TDF model, and there exists a re-definition of the port as well, i.e., there are two branches (one original and one redefined) and both of them connect to the same TDF model, the association is called *PFirm*. Because we are not doing context aware analysis, we do not know which of the two definitions will be used inside the TDF model. Maybe both the definitions are used, or maybe one of them is used. It is context dependent (*if-else, while, for..*), and the condition cannot be evaluated on static time. For e.g., analog mux behavior, (c) if both the branches are redefined and connect to the same TDF model, it is termed as *PWeak*. Because the original port is redefined no matter which path is taken, or (d) if the branches (original and redefined) go to different TDF models, then they are classified according to the individual cases defined above (*Strong or PWeak*).

Coverage Result and Test Adequacy Criteria

Every classification defines a disjoint set of data flow associations. Please note that the strengths of associations are based entirely on static characteristics of the underlying SOC. Hence, a coverage criterion of each classification is required, as follows:

1. *all-Strong*—The criteria are satisfied iff all data flow associations termed *Strong* have been covered.
2. *all-Firm*—All data flow associations termed *Firm* are covered.
3. *all-PFirm*—All data flow associations termed *PFirm* are covered.
4. *all-PWeak*—All data flow associations termed *PWeak* are covered.
5. *all-defs*—Iff at least one def–use association (*v, d, dm, u, um*) is covered for every definition.
6. *all data flow*—Iff all the above criteria are satisfied.

Satisfying *all data flow* criteria is difficult due to limitations of static analysis, i.e., imprecisions or over-approximation. But because the associations are disjoint, satisfaction of each criterion is independent. Hence, it is possible that some of the (sub-)criteria can be fully satisfied—or at least up to a high degree, i.e., 90% of the associations have been exercised. In particular, we expect that *all-Strong*, *all-Firm*, and *all-PFirm* to be the primarily focused criteria. The *Strong*, *Firm*, and PFirm associations contain at least one *du-path*; hence, it is expected from the test input signal to cover them.

Illustration

To illustrate our DFT approach for SystemC-AMS TDF models, we use the example
IOT device given in Fig. 4.7 and its code in Listing 4.7. We show with the help of
three different test-cases applied one at a time how the data flow coverage increases.
The final coverage results are shown in Table 4.7. Table 4.7 can be interpreted in
the following way: the column titled *Static Pairs* lists all the statically identified
data flow associations, and the column titled *Testsuite* presents three test-cases
(TC): *TC1, TC2, TC3*. The table lists the data flow associations from *Strong* (top
left) to *PWeak* (bottom right). If a TCX (X = 1,2,3) is able to exercise a data
flow association, an "x" is marked in the corresponding TC column. The data flow
associations that are not exercised are marked with a "-" in the corresponding TC
column. The test input signals or TC used are: TC1) a constant time continuous
signal of 0.1V, mimicking a temperature of $10\,°C$, TC2) a time continuous signal
from 0V to 0.65V, i.e., $0\,°C$ to $65\,°C$, and back to 0V $(0\,°C)$, TC3) a time continuous
signal at 0.40V $(45\,°C)$. TC1 and TC2 are applied to TS, while TC3 is applied to
HS.

To give an idea, the def–use association *(tmpr, 4, TS, 9, TS)* is Strong as there
exists only one path from Lines 4 to 9, without any re-definition of *tmpr* (a local
variable). The def–use pair is exercised by TC1 and TC2. The def–use pair *(op_intr,
13, TS, 43, ctrl)* is also a Strong association. Because it is not redefined inside the
TDF model *TS*, and being an output port, it is not redefined while flowing to the
next TDF model *ctrl*. The def–use pair *(out_tmpr, 5, TS, 14, TS)* is Firm because it
is local to the TDF model *TS*, and there exists multiple paths from Line 5 to Line
14. The def–use associations *(op_signal_out, 14, TS, 35, AM)* are exercised by both
TC1 and TC2, and *(op_signal_out, 74, sense_top, 36, AM)* is exercised by TC2 only.
There are two paths originating from *op_signal_out* port. One path connects to the
AM as the original signal (defined at Line 14), and the second path is redefined
through a delay element at Line 74. Because both the paths end up in *AM*, any
of them can be used based on the mux *select* line. Since this information cannot be
deduced statically, it is termed PFirm. The def–use pair *(op_mux_out, 77, sense_top,
79, sense_top)* is termed PWeak because the signal at port *op_mux_out* always goes
through the gain element at Line 87; hence, it is redefined before it goes to ADC. It
is exercised by all three test-cases.

When TC2 was applied, it was expected that "T_LED" (Line 49) will be switched
on (once temperature goes above $60\,°C$). But it did not switch on, and the data flow
associations related to lines between Lines 49 and 52 were never exercised. Upon
careful inspection, an interface problem was found between ADC and control. The
output of ADC was saturating because of 9-bit resolution. Any signal above 512 mV
was saturated to 512 mV at ADC output.

TC1 and TC2 alone were not sufficient to achieve a reasonable data flow coverage. They were not able to exercise many associations specific to HS. Hence, TC3
is used additionally. There is still room for coverage improvement. This example
demonstrates that the standard def–use coverage criterion alone is not sufficient
for extensive testing of SystemC-AMS TDF designs. Our proposed SystemC-AMS

Table 4.7 SystemC-AMS TDF models-specific data flow associations—reference Listing 4.7

Static pairs	Testsuite		
	TC1	TC2	TC3
Firm			
(intr_, 6, TS, 13, TS)	–	x	–
(tmp_out, 34, AM, 38, AM)	–	–	–
(out_tmpr, 5, TS, 14, TS)	x	x	–
(intr_, 25, HS, 28, HS)	–	–	x
PFirm			
(op_signal_out, 74, sense_top, 36, AM)	–	x	–
(op_signal_out, 14, TS, 35, AM)	x	x	–
PWeak			
(op_mux_out, 77, sense_top, 79, sense_top)	x	x	x
Strong			
(m_mux_s, 65, ctrl, 66, ctrl)	–	–	x
(m_mux_s, 65, ctrl, 48, ctrl)	–	–	–
(m_mux_s, 65, ctrl, 53, ctrl)	–	–	–
(m_mux_s, 65, ctrl, 61, ctrl)	–	–	x
(m_mux_s, 54, ctrl, 66, ctrl)	–	x	–
(m_mux_s, 54, ctrl, 48, ctrl)	–	x	–
(m_mux_s, 54, ctrl, 53, ctrl)	–	x	–
(m_mux_s, 54, ctrl, 61, ctrl)	–	–	–
(m_mux_s, 59, ctrl, 66, ctrl)	–	x	–
(m_mux_s, 59, ctrl, 48, ctrl)	–	–	–

Static pairs	Testsuite		
	TC1	TC2	TC3
Firm			
(intr_, 27, HS, 28, HS)	–	–	x
(temp, 20, HS, 21, HS)	–	–	x
(newRH, 26, HS, 27, HS)	–	–	x
(newRH, 26, HS, 29, HS)	–	–	x
(Tdepend, 21, HS, 26, HS)	–	–	x
PFirm			
(op_signal_out, 29, HS, 37, AM)	–	–	x
(sensitivity, 24, HS, 26, HS)	–	–	x
(ip_signal_in, 18, HS, 20, HS)	–	–	x
PWeak			
(adc_out, 47, adc, 53, ctrl)	–	x	–
(adc_out, 47, adc, 62, ctrl)	–	–	x
Strong			
(op_intr, 13, TS, 67, ctrl)	x	x	–
(op_hold, 55, ctrl, 7, TS)	–	x	–
(op_hold, 57, ctrl, 7, TS)	–	x	–
(op_clear, 45, ctrl, 8, TS)	x	x	–
(op_clear, 50, ctrl, 8, TS)	–	–	–
(op_clear, 58, ctrl, 8, TS)	–	x	–
(op_clear, 67, ctrl, 8, TS)	x	x	–
(op_hold, 47, ctrl, 7, TS)	x	x	–
(op_hold, 51, ctrl, 7, TS)	–	–	–
(op_intr, 13, TS, 43, ctrl)	x	x	–
(adc_out, 47, adc, 44, ctrl)	x	x	–

(continued)

Table 4.7 (continued)

Static pairs	Testsuite		
	TC1	TC2	TC3
(m_mux_s, 59, ctrl, 53, ctrl)	–	–	–
(m_mux_s, 59, ctrl, 61, ctrl)	–	–	x
(m_mux_s, 63, ctrl, 66, ctrl)	–	–	x
(m_mux_s, 63, ctrl, 48, ctrl)	–	–	–
(m_mux_s, 63, ctrl, 53, ctrl)	–	–	–
(m_mux_s, 63, ctrl, 61, ctrl)	–	–	–
(m_mux_s, 46, ctrl, 66, ctrl)	x	x	–
(m_mux_s, 46, ctrl, 48, ctrl)	–	x	–
(m_mux_s, 46, ctrl, 53, ctrl)	–	x	–
(m_mux_s, 46, ctrl, 61, ctrl)	x	x	–
(m_mux_s, 52, ctrl, 66, ctrl)	–	–	–
(m_mux_s, 52, ctrl, 48, ctrl)	–	–	–
(m_mux_s, 52, ctrl, 53, ctrl)	–	–	–
(m_mux_s, 52, ctrl, 61, ctrl)	–	–	–
(C, 22, HS, 26, HS)	–	–	x
(ip_signal_in, 1, TS, 3, TS)	–	–	–

Static pairs	Testsuite		
	TC1	TC2	TC3
(adc_out, 47, adc, 48, ctrl)	–	x	–
(tmpr, 4, TS, 9, TS)	x	x	–
(op_mux_s, 66, ctrl, 35, AM)	x	x	x
(op_mux_s, 66, ctrl, 36, AM)	–	x	x
(op_mux_s, 66, ctrl, 37, AM)	–	–	x
(sig_in, 3, TS, 4, TS)	x	x	–
(tmpr, 4, TS, 10, TS)	x	x	–
(intr_, 8, TS, 13, TS)	x	x	–
(op_intr, 28, HS, 64, ctrl)	–	–	x
(op_intr, 28, HS, 61, ctrl)	–	–	x
(intr_, 11, TS, 13, TS)	x	x	–
(tmp_out, 35, AM, 38, AM)	x	x	x
(tmp_out, 36, AM, 38, AM)	–	x	–
(tmp_out, 37, AM, 38, AM)	–	–	x
(out_tmpr, 10, TS, 14, TS)	x	x	–
(BC, 23, HS, 26, HS)	–	–	x

TC test-case (test input signal), (x) data flow pair exercised, (–) data flow pair not exercised specific Strong, Firm, PFirm, and PWeak coverage criteria are important for a high quality testsuite.

4.3.3 Implementation Details

In this section, we describe the implementation of our DFT framework. The framework is implemented using the LibTooling library for Clang compiler [83]. Clang generates an Abstract Syntax Tree (AST) of the SystemC-AMS TDF model's source code. The AST is parsed to extract the required information to perform the data flow analysis. We next discuss important implementation details.

The connectivity (binding) of TDF models is possible in different ways depending on the implementation of the SystemC-AMS design, for e.g., using *bind* keyword. It is important to extract it knowing the correct semantics. By default, the implementation of SystemC-AMS TDF models resides in *module::processing()* function, but it could also be in a user defined function. This is registered in the elaboration phase using *register_processing()* library function. This information is required for proper analysis, as the framework needs to know where to look for the TDF model. Afterward, the AST of each TDF model is parsed again, while the signal definition and use information to perform static analysis are extracted. Static analysis is performed in two steps: (1) analysis within a TDF model, (2) analysis of the TDF cluster. Please note, in step one, the output ports defined in a TDF model are assigned X in place of use u, and use model um. In step 2, binding information is used to map the output ports with X to correct TDF models. If the use exists inside the TDF model, the X is replaced with the new use location. Otherwise, it is left as it is. Similarly, the input ports are assigned the start location of their TDF model initially, or location of *initialize()* function. Later, they are also resolved using binding information.

Finally, dynamic analysis is executed by instrumenting the TDF model while parsing AST. For every definition–use detected, a print instruction is written just before the statement. This print instruction logs the information related to the variable/port (location, TDF model name). But this only applies to the statements inside a TDF model. As soon as TDF cluster is analyzed, a print instruction has to be placed either inside the TDF component (could be a library component) or inserted in parallel as a separate TDF model. By parallel insertion, it means that the data (signal) flowing into the re-definition element (gain, delay etc.) also flows into the parallel TDF model, termed *parallel_print()*. We found *parallel_print()* to be less intrusive as the library components remain unchanged. Once the TDF cluster is instrumented, a testsuite is executed. Each test-case (test input signal) exercises different definitions and uses, and data flow associations can be created using the following way. Each definition is mapped on to a corresponding use as soon as it is encountered. If there exists a use, but not definition, it is notified as a warning. The data flow associations are reported to the verification engineer.

Table 4.8 Case study: car window lifter system and Buck-boost converter data flow associations

AMS systems	Iter.	Tests	Static (#)	Data flow associations				
				Dynamic def–use pairs				
				T (#)	S (%)	F (%)	PF (%)	PW (%)
Car window lifter	0	17	573	446	86	81	0	67
	1	20		467	87	82	0	76
	2	23		487	90	84	0	81
	3	26		525	93	89	0	93
Buck-boost converter	0	10	362	243	70	65	100	100
	1	15		268	76	72	100	100
	2	20		282	81	75	100	100
	3	24		307	89	81	100	100

T total, *S* strong, *F* firm, *PF* PFirm, *PW* PWeak

4.3.4 Experimental Results

In this section, we present a case study to demonstrate the proposed DFT approach for SystemC-AMS TDF models. The experiments were carried out on two real-world AMS systems: (1) Car window lifter system, (2) Buck-boost converter [87]. AMS system design implementation and simulations were carried out in COSIDE SystemC-AMS tool environment [115]. Both AMS systems are implemented using SystemC-AMS TDF models. The results are summarized in Table 4.8. In the following, both experiments are briefly explained.

Car Window Lifter System

In the first experiment, we consider a windows lifter system for cars. The AMS system controls the windows movements (up and down), while ensuring the passengers are not harmed. Current flowing through the lifter motor is measured continuously as the window moves. In case of an obstacle (e.g., passenger's hand), the current flow changes, signaling the controller to stop. The system contains an Electrical Control Unit (ECU), and a complete window environment containing the motor, the mechanical parts including the window, and the control buttons. The ECU model includes a motor current filter to remove noise from current measurement, ADC for the motor current conversion, a current detector for over-current detection, the button logic (updown_decoder), and the microcontroller. During the simulation, the obstacle is inserted (and removed) at different times and different window positions into the system.

The car window lifter system has 17 test-cases in the testbench initially, achieving 78% data flow coverage. There were 573 def–use pairs identified by our static analysis, out of which 446 were exercised by the dynamic analysis. Out of 446 exercised pairs, 86% Strong, 81% Firm, and 67% PWeak def–use pairs were exercised. There were no PFirm def–use pairs identified. The *all-defs, all-uses* criteria are not satisfied; hence, *all data flow* is also not satisfied. Table 4.8 shows four iterations where 9 test-cases were added, and coverage increased. During analysis, two types of bugs were discovered: (1) use of ports in TDF models without definitions, (2) dynamic TDF induced failures. In this case, the timestep was reduced to accurately determine the hindrance while closing the window. Due to the change, the threshold comparisons failed in certain cases (specially current feedback loop) leading to def–use pairs being not exercised. These sorts of bugs can prove fatal when undefined behavior goes unchecked.

Buck-Boost Converter

In the second experiment, an energy efficient buck-boost converter [87] is used. It is a type of a DC/DC converter and can operate in two modes: (1) step-down converter (buck) and (2) step-up converter (boost). It is commonly used in IOT devices that are often powered with a battery. The main challenge of buck-boost converter is the switching frequency control algorithm. The algorithm monitors the current flowing. The controller sets the mode of the converter (buck or boost), the expected output value, and the maximum current allowed to flow through the converter. To test the buck-boost converter model, it is checked how fast the expected output voltage is reached and how stable it is. Therefore, an input voltage is applied, and a target voltage is programmed via the controller.

The buck-boost converter has 10 test-cases in the testbench initially, achieving 67% data flow coverage. The static analysis found 362 data flow associations in total, out of which only 243 were exercised by the test-cases. The *all-defs* criteria are not satisfied; hence, *all data flow* is also not satisfied. Out of 243 exercised pairs, 70% Strong, 65% firm, 100% PFirm, and 100% PWeak def–use pairs were exercised. *all-PFirm* and *all-PWeak* criteria are satisfied. Table 4.8 shows four iterations with addition of 14 more test-cases, which increase coverage. We found that in some cases, the ports were not defined but still used in a different TDF model. This is undefined behavior according to SystemC-AMS standards [13]. This cannot be detected by line coverage, as it will still be satisfied.

4.4 Summary

In this chapter, we proposed three novel approaches to strongly enhance the modern VP verification environment. First, we proposed a methodology for SW test qualification of IP integration in a software driven verification flow. Our methodology is based on mutation analysis, and we have shown how to define the main tasks of functional qualification (activate, propagate, detect) in the context of SW test based IP verification. Furthermore, our qualification methodology also relates the coverage results and the SW test results w.r.t. the original and mutated IP blocks. This allows to improve the tests since the user gets information whether for instance a new test is required or the coverage model should be strengthened. We have demonstrated the applicability in a real-world VP showing the integration of two IP blocks.

Additionally, we presented the first DFT approach for SystemC-based VPs and SystemC specific coverage criteria. The criteria use five classifications (Strong, TFirm, TWeak, SyncStrong, SyncWeak) for data flow associations. Furthermore, we explain how to automatically compute the data flow coverage results for a DUV using a combination of static and dynamic analyses. This allows to improve the coverage results by adding tests for uncovered def–use pairs since the users can get useful information. We have demonstrated the applicability in a real-world VP showing the results of one IP model.

Then, we presented the first DFT approach for SystemC-AMS TDF models and a coverage criterion specific to it. At the heart of the proposed work is a scalable static analysis that operates directly on the SystemC-AMS TDF models. It considers the semantics of TDF signal flow and dynamic TDF. Four data flow associations w.r.t. semantics of TDF models are proposed (Strong, Firm, PFirm, PWeak). In addition, using static and dynamic analyses, automatic computation of data flow coverage results is explained for a given DUV. Improvement of coverage results by adding tests for uncovered data flow pairs is also discussed. We have demonstrated the applicability in a real-world VP showing the results of two IP models.

Chapter 5
AMS Enhanced Functional Coverage Verification Environment

This chapter introduces novel systematic verification methodologies leveraging functional coverage to enhance the verification quality of modern VP verification flow. As motivated earlier, the verification of AMS VPs is done by using a signal generator (VP testsuite) on the input side of the VP and assertions/checkers on the output side as shown in Fig. 5.1 in dark gray area. The signal generator creates the input test-stimuli according to the selected static parameters.[1] The assertions/checkers verify the correctness of the functionality. For a thorough verification of the DUV however, tracking of verification progress is required in addition. In digital design verification, coverage is the metric widely used for this purpose. In particular, functional coverage has proven to be very effective [36, 93, 96, 114]. It allows to measure if all the input and output specifications (coverage goals) of the DUV have been verified. Coverage closure is the process used for this task, i.e., to find coverage holes and create new test-stimuli to hit them. In AMS however, verification using functional coverage is still in its infancy [37, 51, 54] as continuous signals and their change over time are much harder to capture. Nevertheless, methodologies driven by functional coverage have also been considered for AMS [14, 106, 107]. However, the existing approaches for AMS suffer from three major shortcomings: (1) they are not systematic, (2) they only consider linear behaviors of AMS systems, and (3) complex mathematical models are required to capture specifications. This is clearly a problem as the true complexity and corner-cases stem from *non-linear* and *unstable* (*overshooting and undershooting*) behaviors. Moreover, capturing these behaviors is non-trivial and requires a lot of effort.

Therefore, this chapter extends the VP verification environment by several components as shown in the yellow area (Fig. 5.1). Coverage bins are introduced on the input–output side of the DUV to capture the specifications. Additionally, novel coverage analysis is used to check for coverage holes. Finally, an automated analysis

---

[1] Static parameters are the parameters that remain constant during the execution of one stimulus signal, e.g., amplitude, frequency, phase, etc.

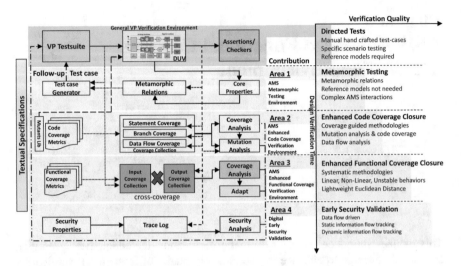

Fig. 5.1 Enhanced functional coverage verification

is proposed to achieve coverage closure. They form the basis for the proposed AMS functional coverage-driven verification approach as detailed in the next sections.

We start in Sect. 5.1 with a brief introduction to functional coverage and outline the deficiencies of AMS VP verification environment. Then in Sect. 5.2, the functional coverage-driven environment is setup, i.e., running example and the ingredients are discussed.

Then, Sect. 5.3 proposes the first functional coverage-driven verification approach as a systematic solution for *Radio Frequency* (RF) amplifiers [*Power Amplifiers (PAs), Low Noise Amplifiers (LNAs), Driver Amplifiers (DAs)*, etc.]. We elevate the main concepts of digital functional coverage to the context of SystemC AMS in particular and system-level simulations in general. First, to enable AMS functional coverage-driven verification, we introduce two coverage refinement parameters on input and output side, to systematically define the input stimuli and capture the DUV specifications. More precisely, on the input side, we define the static parameters[2] of the input stimuli signals using the refinement parameters, i.e., range and input resolution. On the output side, the refinement parameters are used to define the functional coverage to capture the DUV specifications. Second, we present a complete functional coverage-driven characterization approach. At the heart of the approach is the coverage analysis that uses the functional coverage of input, output, and checkers, to determine whether all DUV features according to the specification have been verified. In case of uncovered features, it provides hints to revise the refinement parameters to increase coverage, hence, eventually fully

---

[2] Static parameters are the parameters that remain constant during the execution of one stimuli signal, e.g., amplitude, frequency, phase, etc.

characterizing the DUV. We use an industrial RF transmitter and receiver model as a case study to demonstrate the applicability and efficacy of our approach. This approach has been published in [59].

Next, in Sect. 5.4, we propose a *Lightweight Coverage Directed Stimuli Generation* (LCDG) approach that expands CDG to verify the *linear, non-linear*, and *unstable* behaviors of RF amplifiers at system level. CDG is a verification methodology that aims to reach coverage closure automatically by using coverage data and mathematical functions to direct the next iterations of test-stimuli generation. At the heart of our LCDG approach is a coverage analysis that leverages functional coverage data of input, output, and checkers, in combination with *Euclidean Distance* to achieve coverage closure. The much simpler *Euclidean Distance* in contrast to *Bayesian Networks* or complex probability distribution functions makes our approach *lightweight*. In case of coverage holes, the analysis automatically refines the static stimuli parameters to increase the coverage of the DUV. First, according to the specifications, we define the static parameters of the input stimuli as well as the input–output coverage goals. These coverage goals capture the *linear* and *non-linear* behaviors. While well-known, a lot of corner-cases are hidden in regions when behavior transitions from *linear* to *non-linear* or vice versa. Hence, with our approach, we put a particular focus on these regions since identification of such *unstable* behavior late in the verification phase will have a serious impact on the design. Therefore, second, we define additional dedicated coverage goals to capture *unstable* behavior as well. Note, this definition via coverage goals enables identification of unwanted, i.e., unstable, behavior during simulation. In other words, we are defining here the goals that should not be hit in case of a correct DUV. To define the coverage goals for *linear* and *non-linear* behaviors, we revisit the coverage refinement parameters from Sect. 5.2.2, i.e., *range and resolution*. Additionally, we extend the refinement parameters to define coverage goals for *unstable* behaviors. Third, this allows the following overall LCDG approach: *Euclidean distance* is measured between the coverage holes and the DUV outputs. The DUV output with the smallest *Euclidean distance* to a coverage hole is used to systematically adjust the *resolution* parameter to generate new static parameters for stimuli. This lightweight and systematic approach ensures efficient convergence. In an extensive set of experiments on several configurable industrial system-level LNA models (used as a representative of RF amplifiers), the proposed LCDG approach found a serious bug that escaped during the regular verification process. Furthermore, we perform a fault-injection technique on the industrial LNA to demonstrate the fault detection quality of our LCDG approach. The encouraging results show that LCDG can be effectively used to verify the *linear, non-linear*, and *unstable* behaviors of the RF amplifiers early-on in the design phase.

5.1 Preliminaries

In this section, we briefly review functional coverage concepts. Then we review the typical (advanced industrial) verification environment at the system level and identify its deficiencies.

5.1.1 Functional Coverage

Functional coverage is a verification metric heavily used in digital verification [96]. It determines the extent to which the functionality (or features) of the DUV has been exercised by the input stimuli. Functional coverage is defined by the verification engineer in accordance with the specifications of the DUV. According to the IEEE SystemVerilog standard [72], the following ingredients are used in a simulation-based verification setting. A coverage model is required to define functional coverage, defined as *covergroup*. Each *covergroup* can contain one or more *coverpoints*. Functional coverage maps each functional aspect (or specification) to a *coverpoint*. Each *coverpoint* contains *coverage bins* (sometimes referred to as *bins*) that collect and calculate the number of occurrences of various values. Functional coverage also allows to track information that is received simultaneously on multiple *coverpoints*, called *cross-coverage*. One of the advantages of functional coverage is that it can be reused for verification at different design abstractions.

5.1.2 AMS VP Verification Environment and Deficiencies

To verify an AMS DUV, a verification environment has to be created. Figure 5.2 shows such a verification environment surrounding our running LNA example in dark gray area. Based on this verification environment, the verification engineer performs the verification of the LNA. However, it suffers from two major problems:

1. For a thorough verification of the LNA however, a systematic approach as motivated in the introduction is necessary, which allows to control (and check) the stimuli side as well as to check whether all design features have been characterized. Hence, we extend the verification environment by several components as shown in the light blue area of Fig. 5.2. They form the basis for our proposed AMS functional coverage-driven verification approach as detailed in Sect. 5.3.
2. Even with a systematic approach, manual observation of *non-linear* and *unstable* behaviors becomes a tedious task. Furthermore, capturing such behaviors for automated analysis is not trivial and requires significant effort. Hence, this chapter adds more components to capture such behaviors as shown in the yellow

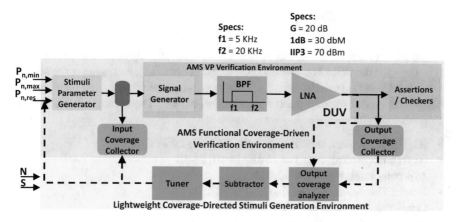

Fig. 5.2 AMS enhanced functional coverage verification environment—LNA: Low Noise Amplifier, BPF: Band-Pass Filter, res: resolution, ires: interval resolution

area of Fig. 5.2. They form the basis for our proposed *Lightweight CDG* (LCDG) approach as detailed in Sect. 5.4.

5.2 Enhanced Functional Coverage Verification Environment Setup

This section describes the ingredients of enhanced functional coverage verification environment for RF amplifiers (see Fig. 5.2). Additionally, it introduces the running example of a *Low Noise Amplifier* (LNA) (Sect. 5.2.1) that will be used in this chapter. There exist four common ingredients to enable the enhanced approach: (1) input stimuli generation, (2) output coverage definition and collection, (3) cross-coverage definition, and (4) checkers definition. Input stimuli generation comprises the following three stages, stimuli static parameter generation, input coverage definition and collection, and signal generation. Input stimuli static parameters are generated and stored in *parameters database*. Different static parameters configure the *signal generator* to generate different input stimuli signals for the DUV, leading the stimulus to exercise different functionality. This can be observed using the AMS functional coverage coverpoints on input and output, i.e., *input coverage collector* and *output coverage collector*. A *cross-coverage* is defined between *input coverage collector* and *output coverage collector* to examine the relationship between inputs and outputs. The *checkers* are used to ensure the correctness of the DUV output behavior. When all the parameters from the *parameters database* have been used and the database is empty, *coverage analysis* is done. The following sections provide more details on the elements.

5.2.1   Running Example: LNA

As running example, we consider two LNAs with slightly different output *Gain* (G). Gain for the first LNA is termed G_1 (used in Sect. 5.3), and gain for the second LNA is termed G_2 (used in Sect. 5.4). The remaining specifications are similar. The behavioral model follows the concepts from [19, 81]. The concrete LNA has the following specifications:

- (G_1) (min., typical, max.) = 16.5 dB, 18.2 dB, 19.9 dB
- (G_2) (min., typical, max.) = 16 dB, 18 dB, 20 dB
- 1dB compression point = 30 dBm
- Output *Third-Order Intercept* (IP3) = 70 dBm
- Operating frequency = 5 KHz to 20 KHz
- Input impedance = 50 Ohms
- Output impedance = 50 Ohms

The model is implemented in SystemC AMS.

5.2.2   Environment Setup

This section describes the components that surround the DUV (running example) to create a verification environment. This environment enables the enhanced functional coverage closure methodologies.

Input Stimuli Generation and Coverage

The first step for AMS functional coverage-driven characterization is the generation of input stimuli signals. They are generated using an ideal signal generator that is defined as a function

$$f(t, p1, p2, p3, \ldots ..pi),$$

where t is the time, and $p1, p2, \ldots, pi;\ 0 < i < \infty$ are the stimuli static parameters that shape the input stimuli signals. The selection of function $f(t, \ldots)$ is based on the DUV functionality to verify, e.g., sine, square wave, single-tone, multi-tone, user defined, etc.

Input stimuli generation comprises three stages, input stimuli static parameter generation, input coverage collection, and signal generation. In the following, these three stages are discussed in detail.

Stimuli Static Parameter Generation

Stimuli static parameters are defined after consulting the specifications of the DUV from the datasheet. The datasheet defines the possible range of input stimuli static parameters, e.g., minimum/maximum frequency (Hz), minimum/maximum amplitude (V), input power range (dBm), etc. Furthermore, an efficient parameter sweeping is required in the defined range, as it is not feasible to simulate for each possible value. Therefore, a *stimuli parameter generator* (Fig. 5.2) is defined, which takes as input a range of the stimuli parameters and an input *resolution* parameter, i.e., the step-size, to systematically generate static parameters. As an example, Fig. 5.3a shows sine waves with *resolution* 0.2 V amplitude difference, and Fig. 5.3b shows sine waves with 0.2 Hz frequency difference. This input *resolution* definition is important, and its selection for stimuli parameters should be carefully done. Too coarse *resolution* will result in having few stimuli static parameters, leading to few input stimuli signals. On the other hand, too fine *resolution* will result in too many input stimuli signals. In case of former, the simulation time is reduced at the expense of possibly unverified DUV, whereas, for the latter, the simulation time is extremely high and possibly a fully verified DUV. Please note that coarse *resolution* means a bigger step-size, and a fine *resolution* means a smaller step-size. The *stimuli parameter generator* generates all possible combinations of static parameters w.r.t. the defined *resolution* and stores them in the *parameters database*.

For our running example of LNA, we take an ideal signal generator that takes two arguments as input and has the following function implemented:

$$f_1(t, A, F) = A * sin(2 \times \pi \times F \times t), \tag{5.1}$$

where parameter A is the amplitude, parameter F is the frequency, and parameter t is the time of the stimuli signal. A and F are static parameters that need to be defined. Hence, from the specifications of (Sect. 5.2.1), initially, we set the inputs of the stimuli parameter generator as follows:

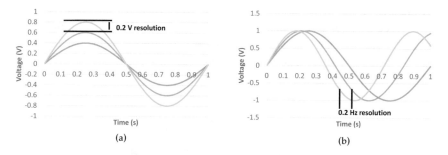

Fig. 5.3 Graphical illustration of *resolution* parameter. (**a**) Stimuli parameter resolution definition with 0.2 V amplitude difference. (**b**) Stimuli parameter resolution definition with 0.2 Hz frequency difference

$$A = (0, 5.0; 0.5)\ V$$
$$F = (5e3, 20e3; 3e3)\ Hz. \tag{5.2}$$

Equation 5.2 specifies parameter A: between 0 V and 5 V (both inclusive) and the *resolution* is 0.5 V. It means the amplitude values should be 0 V, 0.5 V, 1 V,...Eq. 5.2 also specifies parameter F: between 5 KHz and 20 KHz (both inclusive) and the *resolution* is 3 KHz. It means the frequency values should be 5 KHz, 8 KHz, ... The *stimuli parameter generator* takes these two parameters as inputs and generates pairs of parameters (a, f). The concrete pairs are: (a, f) = 0, 5e3; 0.5, 5e3; 1, 5e3;...; 4, 20e3, ...

Input Coverage Collection

Functional coverage coverpoints are required on the input side of the DUV to assess the quality of the generated static parameters. Hence, an *input coverage collector* is defined and placed before the *signal generator*. The reason for this placement is that the static parameters are directly available. Otherwise, if placement is done after *signal generator*, then complex measurement tools and post-processing of the signal are required to extract the same parameters. The *input coverage collector* captures different stimuli static parameters, e.g., frequency (Hz), amplitude (V), etc. Additionally, it captures input specifications of the DUV, e.g., input power (dBm). The functional coverage *bins* for static parameters and input specifications are defined using the range of static parameters. Moreover, each *bin* defines another parameter *interval resolution*. Its purpose is to create a range of values inside the defined *bin resolution*. Please note that it is quite possible that the selected *resolution* and *interval resolution* result in *coverage bins* that do not cover all DUV input specifications. In this case, a revision of *resolution* and *interval resolution* is required. The *input coverage collector* plays a vital role in cross-coverage, explained later.

A relevant code snippet of the running example Sect. 5.2.1 for input coverage *coverpoints* is shown in Listing 5.1. In the input coverage collector, *coverage bins* are defined for A (Listing 5.1 Lines 1–7), and F (Listing 5.1 Lines 9–15) using initially the same *resolution* as defined for stimuli static parameters in Eq. 5.2. Hence, each *bin* represents one amplitude/frequency value. One additional *coverpoint* is defined for input power as follows:

$$input\_power = (20, 30; 5, 4.99)\ dBm$$
$$input\_power = (30, 36; 2, 1.99)\ dBm \tag{5.3}$$
$$input\_power = (38, 40; 1, 0.99)\ dBm$$

Equation 5.3 defines coverage *bins* for input power with three different *resolution* and *interval resolution* values: (1) between 20 dBm and 30 dBm with 5 dBm *resolution* and 4.99 dBm *interval resolution*, (2) between 30 dBm and 36 dBm with 2 dBm *resolution* and 1.99 dBm *interval resolution*, and (3) between 38 dBm and 40 dBm with 1 dBm *resolution* and 0.99 dBm *interval resolution*. This is because of the logarithmic scale for dBm where the increase in amplitude (A) of the stimuli signals results in smaller increases in dBm. Hence, these three different values are used to cover maximum behavior of the DUV. The *bins* are defined from Listing 5.1 Line 18 to Listing 5.1 Line 26. The defined *bins* for *A, F, and input power* are shown in Tables 5.1, 5.2, and 5.3, respectively. The tables can be interpreted in the following way, the first column shows the overall coverage of the *coverpoint*, the second column shows the description of the *coverage bins*, the third column shows the value of each *bin* that should appear on the output of DUV for this *bin* to be covered, the fourth column shows the number of times this *bin* was covered (hit), and the last column shows if the *bin* was covered (hit) or not. *Green* color reflects a successful hit, and *red* color reflects a miss.

```
1    coverpoint<double> A_cvp = coverpoint<double> (this,
2    bin<double>("low corner", 0.0),
3    bin<double>("0.5", 0.5),
4    ....
5    bin<double>("4.5", 4.5),
6    bin<double>("high corner", 5.0)
7    ) ;
8
9    coverpoint<double> F_cvp = coverpoint<double> (this,
10   bin<double>("low corner", 5e3),
11   bin<double>("8e3", 8e3),
12   ....
13   bin<double>("17e3", 17e3),
14   bin<double>("high corner", 20e3)
15   ) ;
16
17   coverpoint<double> input_power_cvp = coverpoint<double> (this,
18   bin<double>("20", interval(20:24.99)), // 4.99 dBm
19   bin<double>("25", interval(25:29.99)),
20   bin<double>("30", interval(30:31.99)), // 1.99 dBm
21   bin<double>("32", interval(32:33.99)),
22   bin<double>("34", interval(34:35.99)),
23   bin<double>("36", interval(36:37.99)),
24   bin<double>("38", interval(38:38.99)), // 0.99 dBm
25   bin<double>("39", interval(39:39.99)),
26   bin<double>("40", interval(40:40.99))
27   ) ;
```

Listing 5.1 LNA: input coverage coverpoints definition

Signal Generation

The stimuli static parameters stored in the *parameter database* (Sect. 5.2.2) are taken out one pair at a time and given as input to the *signal generator*. The *signal generator* generates the corresponding test input signal for the DUV. When the stimuli parameters are applied, we observe that the input coverage is 100% for amplitude A, frequency F, and input power (dBm) as shown in Tables 5.1, 5.2, and 5.3, respectively.

Table 5.1 LNA: Parameter A (amplitude) coverage report

	amplitude_cvp "amplitude_values"			
	Description	Value	#Hits	Hit
100.0%	Low corner	0	18	✓
	0.5 V	0.5	18	✓
	1.0 V	1.0	18	✓
	1.5 V	1.5	18	✓
	2.0 V	2.0	18	✓
	2.5 V	2.5	18	✓
	3.0 V	3.0	18	✓
	3.5 V	3.5	18	✓
	4.0 V	4.0	18	✓
	4.5 V	4.5	18	✓
	High corner	5.0	18	✓

Resolution = 0.5 V

Table 5.2 LNA: Parameter F (frequency) coverage report

	freq_cvp "frequency_values"			
	Description	Value	#Hits	Hit
100.0%	Low corner	5000	33	✓
	8 KHz	8000	33	✓
	11 KHz	11,000	33	✓
	14 KHz	14,000	33	✓
	17 KHz	17,000	33	✓
	High corner	20,000	18	✓

Resolution = 3 KHz

Table 5.3 LNA: Input power coverage report

	input_power_cvp "input_power_values"			
	Description	Value	#Hits	Hit
100.0%	20 dBm	[20: 24.99]	18	✓
	25 dBm	[25: 29.99]	18	✓
	30 dBm	[30: 31.99]	18	✓
	32 dBm	[32: 33.99]	18	✓
	34 dBm	[34: 35.99]	18	✓
	36 dBm	[36: 37.99]	18	✓
	38 dBm	[38: 38.99]	18	✓
	39 dBm	[39: 39.99]	18	✓
	40 dBm	[40: 40.99]	18	✓

Resolution = 5, 2, 1 (dBm), *Interval resolution* = 4.99, 1.99, 0.99 (dBm)

Output Coverage Definition and Collection

The input stimuli signals generated by *signal generator* exercise different DUV behaviors. The output of DUV goes to *output coverage collector*. The *output coverage collector* is defined to capture different DUV specifications, i.e., output

Table 5.4 LNA gain (G_1) output coverage report

	Description	Value	#Hits	Hit
	gain_cvp "gain_values"			
37.5%	Low corner	[16.5: 16.7]	0	x
	17.0 dB	[17.0: 17.2]	0	x
	17.5 dB	[17.5: 17.7]	0	x
	18.0 dB	[18.0: 18.2]	21	✓
	18.5 dB	[18.5: 18.7]	0	x
	19.0 dB	[19.0: 19.2]	27	✓
	19.5 dB	[19.5: 19.7]	21	✓
	High corner	[20.0: 20.2]	0	x

Resolution = 0.5 dB, *interval resolution* = 0.2 dB

signal power (dBm), gain (dB), 1 dB compression point, etc. Again, the *resolution* of output coverage collector is defined. Too coarse (low *resolution*) may miss important specifications, creating coverage holes. Too fine (high *resolution*) may lead to too many unnecessary values exhibiting similar behavior. Furthermore, *interval resolution* similar to Sect. 5.2.2 is also defined. The bins with more hits can further be expanded to observe exact behavior.

The output specifications of interest for our running example of LNA DUV are gain (G_1) and 1 dB compression point. We define a coverpoint for gain with *bins* ranging from 16.5 dB to 20 dB, with a *resolution* of 0.5 dB and an *interval resolution* of 0.2 dB. The 1 dB compression point lies at the input power corresponding to 18.9 dB gain. The coverage bins are shown in Table 5.4.

Cross-Coverage Definition

Cross-coverage is required to observe the input–output relationship of the DUV. Since the goal is to have an early verification, cross-coverage information can exactly pin-point which input stimuli exercised what output behavior. This is useful in particular because the same stimuli are considered for lower abstractions such as SPICE-level simulations. Instead of executing every test-stimuli that will take significant time, only useful stimuli can be executed to get the desired DUV behavior. Table 5.7 shows a snippet of the cross-coverage between checkers—input power—gain.

Checkers Definition

Functional coverage only tells how much functionality of the DUV has been exercised. In order to verify if the covered functionality also exhibited correct behavior, checkers are required. Hence, DUV output is also used as input for checkers as shown in Fig. 5.2. A checker definition for calculating 1 dB compression point is shown in Listing 5.2 for reference. It checks that the DUV output signal stays within a ± epsilon range (Line 12). When the DUV output is multiplied with a factor of 0.891, it gives us the amplitude value of 1 dB compression point [81, 116]. Line 1 defines a reference signal *f_inp1* that is compared to the DUV output (Line 10) using the checkers. The comparison starts after 2 ms (Line 13) so that the signal has stabilized and continues until 10 ms (Line 14). If the checker fails at any time, a warning message is generated (Line 15).

```
1    auto f_inp1 = [=](double t) { ()
2    return dut_p->input_amplitude*sin(
3    2.0*M_PI*dut_p->input_frequency*t
4    );
5    };
6
7    CHECK_RANGE_EPS_DYNAMIC(*s.lna_out_s, // Checked Signal
8    ref_f_t([=](double t) //reference (expected) function
9    {
10   return ( 10*0.891*f_inp1(t)); ()
11   }),
12   0.2, // epsilon ()
13   sc_time(2.0, SC_MS), ()
14   sc_time(10, SC_MS), ()
15   sc_core::SC_WARNING, ()
16   "top_testbench");
```

Listing 5.2 LNA: checker definition for calculating 1 dB compression point

In the next section (Sect. 5.3), we use the ingredients to propose the functional coverage-driven verification approach to enhance the verification quality of RF amplifiers systematically.

5.3   AMS Functional Coverage-Driven Verification Approach

In this section, we discuss systematic functional coverage-driven verification approach. The overview of the approach is shown in Fig. 5.4. The approach uses the main ingredients from Sect. 5.2: input stimuli generation, output coverage definition, cross-coverage definition, and checkers definition. The ingredients are used in the novel coverage analysis proposed in Sect. 5.3.1.

5.3.1   Coverage Analysis

The coverage analysis is executed when all stimuli signals corresponding to static parameters have been generated and the *parameters database* is empty. The generated coverage results are analyzed to determine if all input *coverage bins* and output *coverage bins* have been individually covered or not. The input coverage should be 100% because it depends on the static stimuli parameters and not on

Fig. 5.4 Overview on AMS functional coverage-driven approach—res: resolution, ires: interval resolution

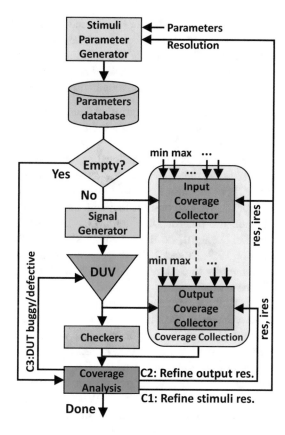

DUV's behavior. It is evident from the results in Tables 5.1, 5.2, and 5.3. The output functional coverage is dependent on the input stimuli signals and DUV's behavior. Our goal is to achieve 100% output coverage. If all the coverage bins are covered in the coverage report, i.e., 100% coverage, nothing else is required to be done. Please note that 100% coverage only indicates that all the defined objectives (*coverage bins*) of the DUV characterization have been achieved. It does not necessarily mean that all DUV specifications have been verified. In case the DUV specification is not covered, there are three cases:

C1: The *resolution* of the static stimuli parameters is coarse (low); hence, the input stimuli signals are far apart. As a consequence, many output *coverage bins* are not hit, and coverage holes appear. In this case, the *resolution* should be increased, i.e., step-size should be decreased.

C2: The *resolution* of the *output coverage collectors* is coarse (low) or the *bins* are not defined. Hence, it is not possible to capture DUV's output specifications.

In this case, either new *bins* need to be defined (as per specifications), or the *resolution* needs to be refined, i.e., decrease step-size.

C3: The DUV has a bug, and refining *resolution* or *interval resolution* of static stimuli parameters or *output coverage collectors* has no effect. In this case, analyze the DUV implementation for possible bugs.

Case C2 has a higher priority than case C1. The reason is that if *bins* for a specification are not defined in the *output coverage collector*, the output behavior cannot be observed.

We observe that the functional coverage for gain (G_1) in our running example is not 100%, rather only 37.5% as shown in Table 5.4. According to Sect. 5.3.1 case C1, to increase the output coverage, we need to refine the stimuli parameter resolution. We also observe that with the defined *resolution*, and *interval resolution* of gain, the bin for DUV gain, i.e., 19.9 dB, and the 1 dB compression point bin, i.e., 18.9 dB, do not appear as shown in Table 5.4. Please note that even if gain of 19.9 dB and 1 dB compression point were achieved in functionality, we have no way to capture it. Therefore, we have coverage holes. According to case C2, new bins need to be defined. For C2, we add new bins ranging from 18.8 dB to 19.9 dB with a resolution of 0.2 dB and an *interval resolution* of 0.1 dB. But we observe that the bins are still uncovered with the functional coverage of 41.66% (Table 5.5). As per case C1, we increase the *resolution* (decrease step-size) of A from 0.5 V to 0.2 V, and F from 3 KHz to 1 KHz. We observe that the output coverage increases to 46%. We further refine the resolution of stimuli parameter A to 0.1 V, but we do not observe any improvement in coverage. As per case C3, we analyze the DUV's implementation and find a bug. Figure 5.5 shows the gain of the buggy DUV (actual gain) and bug-free DUV (expected gain), where the LNA does not saturate with increasing input power. The gain is only

Table 5.5 LNA gain (G_1) coverage report. Case C2: addition of bins in gain coverpoint between 18.8 dB and 19.9 dB

gain_cvp "gain_values"			
Description	Value	#Hits	Hit
Low corner	[16.5: 16.7]	0	x
17.0 dB	[17.0: 17.2]	0	x
17.5 dB	[17.5: 17.7]	0	x
18.0 dB	[18.0: 18.2]	21	✓
18.5 dB	[18.5: 18.7]	0	x
18.8 dB	[18.8: 18.9]	0	x
19.0 dB	[19.0: 19.1]	9	✓
19.2 dB	[19.2: 19.3]	0	x
19.4 dB	[19.4: 19.5]	9	✓
19.6 dB	[19.6: 19.7]	9	✓
19.8 dB	[19.8: 19.9]	9	✓
20.0 dB	[20.0: 20.2]	0	x

41.66%

Resolution gain = 0.2 dB, *interval resolution* gain = 0.1 dB

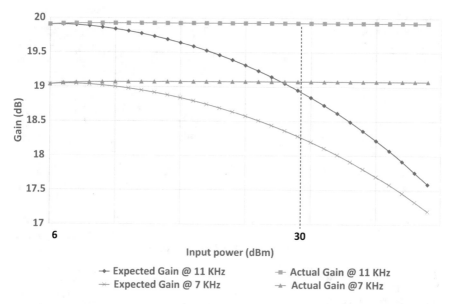

Fig. 5.5 Gain (G_1) vs. input power

shown for 2 input frequencies for ease of understanding. The expected gain is never observed, and hence, the checkers are also never triggered. Once the bug is fixed, the output coverage becomes 100% (Table 5.6), and the checker is triggered to verify the correct behavior. A part of cross-coverage report is displayed in Table 5.7, which shows the cross-coverage between checkers, input power, and gain. The first row of Table 5.7 can be interpreted as follows: the checker indicates the correct functionality of the DUV with *pass*, the input power at the time when the correct functionality was observed was 30 dBm, and the output gain was 18.9 dB. The functional coverage not only shows the exact behavior, but also the complete range of operation of the DUV, which is generally required.

5.3.2 Industrial Case Study

In this section, we present a case study using an industrial RF transmitter and receiver model (Fig. 5.6) to show the efficacy of our proposed approach. The system uses LNAs at different positions in the systems. The AMS model is implemented in SystemC AMS, and the simulations are carried out using the commercial tool COSIDE [115].

The system models a complete RF transmitter and receiver structure. Transmitted symbols are transferred using a *Differential Quadrature Phase Shift Keying* (DQPSK) modulation. After an up-sampling of the encoded signal, the signal

Table 5.6 LNA gain (G_1) coverage report. Case C1: static parameter refinement on input stimuli

	gain_cvp "gain_values"			
	Description	Value	#Hits	Hit
100%	Low corner	[16.5: 16.7]	9	✓
	17.0 dB	[17.0: 17.2]	18	✓
	17.5 dB	[17.5: 17.7]	96	✓
	18.0 dB	[18.0: 18.2]	219	✓
	18.5 dB	[18.5: 18.7]	24	✓
	18.8 dB	[18.8: 18.9]	66	✓
	19.0 dB	[19.0: 19.1]	144	✓
	19.2 dB	[19.2: 19.3]	9	✓
	19.4 dB	[19.4: 19.5]	132	✓
	19.6 dB	[19.6: 19.7]	111	✓
	19.8 dB	[19.8: 19.9]	162	✓
	20.0 dB	[20.0: 20.2]	18	✓

Resolution A $= 0.1$ V, *resolution* F $= 1$ KHz

Table 5.7 Cross-coverage bug-free DUV (excerpt)—checker vs. input power vs. gain (G_1)

	Checker vs. input power vs. gain		
	Description	#Hits	Hit
92.4%	Pass 30 dBm × 18.9 dB	66	✓
	Pass × 32 dBm × 18.8 dB	66	✓
	Pass × 36 dBm × 18.65 dB	26	✓
	Pass × 40 dBm × 18.52 dB	26	✓

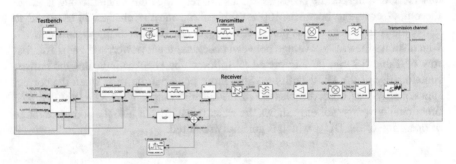

Fig. 5.6 Case study: RF transmitter and receiver model

is mixed up to the transmission frequency. To model the complete transmission chain, an *Additive White Gaussian Noise* (AWGN) channel model is used. Thereby, white Gaussian noise (white noise) is added to the transmitted signal. The receiver first amplifies the received signal using a LNA and then mixes it down from the transmission frequency during demodulation. Afterward, a detector tries to decide which signals have been sent and maps them on the symbols.

For verifying the overall system, a testbench was created, which sends random symbols over the RF transmitter and compares the received symbols on the receiver side. Based on this, a bit error rate can be calculated. The verification goal is to evaluate the quality of the testbench and to make sure all the functionality (w.r.t. specifications) of different models is covered; hence, functional coverage is used. The parameters of interest for our experiment are *gain, IIP3, 1 dB point*. The coverpoints are placed at the following points in the transmitter (Fig. 5.6) at the input and output of: LNA (*i_gain_cplx1*) with gain = 6 dB, IIP3 = 30 dBm, 1 dB point = 180 dBm, and input–output resistance = 100 ohm. In the receiver chain: (1) LNA (*i_lna_base_pb1*) with gain = 20 dB, IIP3 = 20 dBm, 1 dB point = 60.4 dBm, input–output resistance = 50 ohm, (2) LNA (*i_gain_cplx2*) with gain = 20 dB, IIP3 = 10 dBm, 1 dB point = 60.4 dBm, input–output resistance = 100 ohm. Coverpoints were created to check the *gain (G)* (dB), *1 dB compression point* (dBm), and *third-order input intercept point (IIP3)* (dBm) characteristics. The refinement parameters: *resolution* and *interval resolution* for gain are 2 dB and 0.5 dB, respectively, in the first iteration. *Resolution* and *interval resolution* for other characteristics are 10 dBm and 5 dBm, respectively, in the first iteration. Figure 5.7 shows the development of coverage for each LNA over three iterations using the proposed approach. In Fig. 5.7, each LNA model's coverage is shown with different color sets in each iteration.

It was to be expected that not all coverpoints used in block level testbench are covered. However, our approach provides insight on the DUV operation: (1) out of scope of the intended specifications, (2) coverpoints have to be refined with a different *resolution* and *interval resolution*, as they are not detailed enough to capture the use case within the complex system. In each iteration, the *resolution* and *interval resolution* were refined to precisely check the covered characteristics. Interestingly, the gain coverpoints in the transmitter chain behaved as expected.

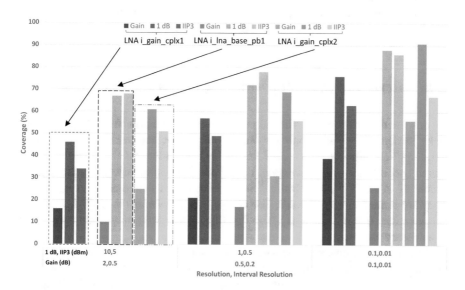

Fig. 5.7 Case study—coverage closure w.r.t. resolution

There was no non-linear behavior; hence, the total coverage remained low. As shown in Fig. 5.7, some of the IIP3 bins were hit, but they were hit in the lower power ranges. Hence, the coverage was always low. But in the receiver chain, non-linearity in the LNAs was observed because of AWGN, and the 1 dB compression point was hit. Some coverage bins for IIP3 point were also covered as shown in Fig. 5.7. Figure 5.7 shows the increasing progression of different characteristics as the refinement parameters, *resolution* and *interval resolution*, are refined over the iterations. In summary, the proposed AMS functional coverage-driven characterization approach for RF amplifiers systematically increased the coverage to verify the RF transmitter and receiver models.

The next section extends these concepts to propose a lightweight CDG approach to verify the *linear*, *non-linear*, and *unstable* behaviors of RF amplifiers.

5.4 Lightweight Coverage Directed Stimuli Generation

In this section, we discuss the proposed LCDG approach to automatically verify the *linear*, *non-linear*, and *unstable* behaviors of RF amplifiers. To enable LCDG approach, the functional coverage-driven verification environment in Fig. 5.2 is extended by several components in yellow area. It comprises automated coverage analysis using *Euclidean distance* and post-processing and static parameters tuning. The approach uses the concepts discussed earlier in Sect. 5.2.2; however, in the next section (Sect. 5.4.1), we revisit the output coverage definition from LCDG perspective. The overview of the overall approach is shown in Fig. 5.8. The approach uses two control parameters, N and S, to control the execution time of the analysis. The lightweight coverage analysis will be discussed in Sect. 5.4.2 in light of Fig. 5.8.

5.4.1 Revisiting Output Coverage Definition and Collection

In this section, we revisit the output coverage definition from Sect. 5.2.2 in order to define coverage goals for *linear*, *non-linear*, and *unstable* behaviors of RF amplifiers. The *coverage bins* are defined according to the range of DUV output specifications and *resolution*. First, *linear* and *non-linear* goals are defined according to the DUV specifications at a resolution and should always be satisfied. They are termed *coverage goals*. Too coarse (low *resolution*) may miss important specifications that are necessary. Too fine (high *resolution*) may lead to too many unnecessary values.

To capture the unstable behavior of RF amplifier, i.e., if the behavior of RF amplifier overshoots or undershoots at some test-stimuli, *unstable coverage goals* are defined. These goals are defined to not be hit when the DUV is functioning properly. These *unstable coverage goals* are further divided into *overshoot/undershoot* (O/U) and *secondary-extension* coverage goals. *O/U* goals are defined by defining 8

Fig. 5.8 Overview on LCDG
approach—res: resolution,
ires: interval resolution,
L/NL: linear/non-linear
coverage goals, O/U:
overshoot/undershoot
coverage goals, SCG:
secondary-extension coverage
goals

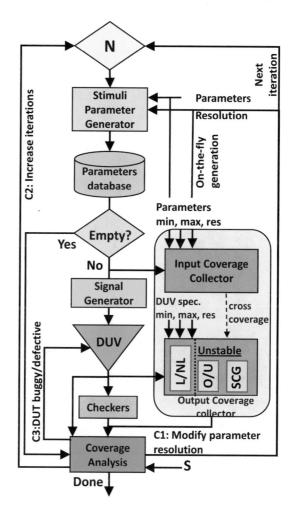

additional goals on both ends of the *coverage goals* with the *resolution* of $\frac{resolution}{4}$.
Furthermore, the *interval resolution* parameter is defined to capture all the possible
values in a certain interval. These intervals provide an overview of the values. Since
we do not know what (unstable) behavior might be observed, an interval provides a
better chance of capturing it. *Secondary-extension* goals are defined by adding bins
between the range of values defined in DUV specifications. It is done by reducing
the *resolution*, i.e., $\frac{resolution}{4}$. Again, the *interval resolution* is defined to create an
interval of values.

To illustrate this, consider the output gain (G_2) for our running example LNA
(Sect. 5.2.1). We define a coverpoint for G_2 with *coverage goals* ranging from 16
dB to 20 dB with a *resolution* of 1 dB. The *coverage goals* are shown in Table 5.8,
and the *unstable coverage goals* are shown in Tables 5.9 and 5.10. Furthermore, the

Table 5.8 LNA gain (G_2) output coverage report—iteration 1

gain_cvp "gain_values" iteration 1			
Description	Value	#Hits	Hit
16.0 dB	16.0	0	x
17.0 dB	17.0	0	x
18.0 dB	18.0	7	✓
19.0 dB	19.0	0	x
20.0 dB	20.0	0	x

20%

Resolution = 1.0 dB

Table 5.9 LNA gain (G_2)—overshoot/undershoot coverage goals

gain_cvp "gain_values" iteration 1			
Description	Value	#Hits	Hit
14.00 dB	[14.00:14.24]	0	x
14.25 dB	[14.25:14.49]	0	x
14.50 dB	[14.50:14.74]	0	x
14.75 dB	[14.75:14.99]	0	x
15.00 dB	[15.00:15.24]	0	x
15.25 dB	[15.25:15.49]	0	x
15.50 dB	[15.50:15.74]	0	x
15.75 dB	[15.75:15.99]	0	x
21.00 dB	[21.00:21.24]	0	x
21.25 dB	[21.25:21.49]	0	x
21.50 dB	[21.50:21.74]	0	x
21.75 dB	[21.75:21.99]	0	x
22.00 dB	[22.00:22.24]	0	x
22.25 dB	[22.25:22.49]	0	x
22.50 dB	[22.50:22.74]	0	x
22.75 dB	[22.75: 22.99]	0	x

0%

Resolution = 0.25 dB, *interval resolution* = 0.24 dB

complete DUV output trace is also saved separately as shown in Table 5.11 (more details in Sect. 5.4.2).

5.4.2   Lightweight Coverage Analysis

The lightweight coverage analysis is executed when all stimuli signals corresponding to the static parameters have been generated, and the *parameters database* is empty (see Fig. 5.8). The generated coverage results are automatically analyzed to determine if all input *coverage bins* and output *coverage goals* have been individually covered or not. The input coverage should be 100% because it depends on the static stimuli parameters and not on DUV's behavior. The goal of the coverage analysis is to achieve 100% *coverage goals* for the specifications of interest. If all the *coverage goals* are hit in the output coverage report, nothing else is

Table 5.10 LNA gain (G_2)—secondary-extension goals

	Description	Value	#Hits	Hit
	gain_cvp "gain_values" iteration 1			
66.6%	16.25 dB	[16.25:16.49]	2	✓
	16.5 dB	[16.5:16.74]	6	✓
	16.75 dB	[16.75:16.99]	6	✓
	17.25 dB	[17.25:17.49]	0	x
	17.5 dB	[17.5:17.74]	21	✓
	17.75 dB	[17.75:17.99]	0	x
	18.25 dB	[18.25:18.49]	4	✓
	18.5 dB	[18.5:18.74]	4	✓
	18.75 dB	[18.75:18.99]	0	x
	19.25 dB	[19.25:19.49]	0	✓
	19.5 dB	[19.5:19.74]	4	✓
	19.75 dB	[19.75:19.99]	0	x

Resolution $= 0.25$ dB, *interval resolution* $= 0.24$ dB

Table 5.11 Excerpt of LNA gain (G_2) output trace after iteration 1

	Description	Value
	DUV gain output trace	
–	16.6 dB	16.6
	16.7 dB	16.7
	17.2 dB	17.2
	18.0 dB	18.0
	18.4 dB	18.4
	18.6 dB	18.6
	19.6 dB	19.6

required to be done,[3] and the coverage analysis switches to *unstable coverage goals*. However, if the *coverage goals* are not covered, there are three cases (Fig. 5.8):

C1: The *resolution* of the static stimuli parameters requires adjustment. As a consequence, many output *coverage bins* are not hit and coverage holes appear. In this case, the *resolution* should be adjusted, i.e., step-size should be increased/decreased.

C2: The number of iterations N are too few. As a consequence, not enough test-stimuli is generated to exercise DUV behavior. In this case, when N iterations are finished, increase them by a factor of 2, i.e., double them.

C3: The DUV specifications have an error, or the DUV has a bug. Refining *resolution* of static stimuli parameters and increasing N have no effect. This only occurs in case of *hard coverage goals*. In this case, analyze the DUV specifications (*coverage goals*) and implementation for possible bugs.

---

[3] Please note that 100% coverage only indicates that all the defined objectives have been achieved. It does not necessarily mean that all DUV specifications have been verified.

These cases are handled using an automated feedback loop that uses the complete DUV output trace as a foundation. The feedback loop consists of three main components: (1) *Output coverage analyzer*, (2) *Subtractor*, (3) and *Tuner*. The pseudocode for the algorithm is shown in Algorithm 1.

It takes static parameters (*params(min, max, res)*), N, S, and output specifications (*output_specs(min,max,res)*) as input and generates coverage report on output when the analysis finishes. During the initialization, output coverage bins are defined along with initial parameters (Line 1). Every time the new static parameters are generated (with new *resolution* Lines 5 and 32), new input coverage bins are dynamically generated in Line 6. The cross-coverage between input and output coverage bins is dynamically defined in each iteration (Line 7). The execution of all static stimuli parameter pairs (stored in *Database*) is done as shown in Lines 8–12. The *SignalGenerator* is configured iteratively with static stimuli parameters and the simulation starts. Afterward, the coverage reports are generated, and the DUV output trace is saved. In the following, we explain the major components of LCDG approach in detail.

Output Coverage Analyzer

If the coverage is not 100%, the analysis starts with case C1. The goal is to find the *resolution* of static stimuli parameters that maximizes the chances of hitting a coverage hole. Hence, the output coverage results of the DUV are analyzed after each iteration. The analysis is executed in two stages: (1) coverage holes detection (Algorithm 1 Line 25), (2) *nearest neighbor* (nn) identification via *Euclidean Distance* (Algorithm 1 Line 30).

Coverage Hole Detection

The first step is to find the coverage holes in the output coverage report. Instead of randomly targeting multiple uncovered events, we systematically search for the first unhit bin and declare it the coverage goal for the next iteration.

Nearest Neighbor Identification via Euclidean Distance

In the second step, we look for the nearest neighbor of the new coverage goal. A nearest neighbor is a value that occurred for which we have evidence [from DUV output trace (Table 5.11)], but this value may not be defined in *coverage goals*. This approach works because we already know that the nearest neighbor occurred (whether defined as coverage goal or not); hence, this information can be used to systematically create a feedback path to fine-tune the static stimuli parameters.

Algorithm 1: Proposed LCDG approach to capture *linear*, *non-linear*, and *unstable* behaviors

---

Data: params(min,max,res), N, S, output_specs(min,max,res)
Result: DUV coverage report

1 out_bins = define_output_coverage_bins(output_specs);
2 local_params = params;
3 **while** *COUNT not reached* **do**
4   **while** *N not reached* **do**
5   Database = GenerateStaticPairs(local_params));
6   in_bins = create_input_coverage_bins(local_params);
7   define_cross_coverage(in_bins, out_bins);
8   **while** *!Database.empty()* **do**
9   SignalGenerator (database.front.params);
10   start_simulation();
11   generate_coverage_report();
12   save_duv_output_trace();

13   duv_trace = get_duv_trace();
14   oc = get_output_coverage();
15   OUc = get_OU_coverage();
16   soc = get_output_secondary_coverage();
17   **if** *oc == 100% && !CoverageDone* **then**
18   CoverageDone = true; reset_N(10);

19   **if** *(OUc == 100% || executed 10 times) && !OUCovDone && CoverageDone* **then**
20   OUCovDone = true; reset_N(10);

21   **if** *(soc == 100% || executed 10 times) && !SecCovDone && CoverageDone &&*
        OUCovDone **then**
22   SecCovDone = true; reset_N(10);

23   // C1: refine resolution
24   **if** *!CoverageDone* **then**
25   coverage_hole = FindCoverageHole(oc);

26   **else if** *!OUCovDone* **then**
27   coverage_hole = FindCoverageHole(OUc);

28   **else if** *!SecCovDone* **then**
29   coverage_hole = FindCoverageHole(soc);

30   nn = NearestNeighbor(duv_trace, coverage_hole);
31   SO = Subtractor (coverage_hole, nn);
32   Tuner (S, SO, local_param);

33   // C2: increase iterations N
34   **if** *!CoverageDone* **then**
35   COUNT++; reset_N(N × 2);

36 **if** *!CoverageDone* **then**
37   // C3: DUV has possibly bug

Therefore, slightly modifying the *resolution* can help achieve the coverage goal.[4]
We take advantage of the *Euclidean distance* induced by the absolute-value norm to
find the neighboring values. We take the absolute of difference between the coverage
goal and the values generated in the DUV output trace at index i, as shown in Eq. 5.4.

$$nn = |\ coverage\_hole - duv\_trace(i)\ |. \tag{5.4}$$

The nearest neighbor is the value with the least Euclidean distance. Once the nearest
neighbor (Line 30) is found, it is forwarded to the *Subtractor* along with the new
coverage goal.

Subtractor

The *Subtractor* module takes the difference between the nearest neighbor and the
coverage hole (Eq. 5.5).

$$sub\_out\ (SO) = coverage\_hole - nn. \tag{5.5}$$

The SO can never be 0 because it signifies verified specification. Therefore, the SO
will be either positive or negative. The SO calculation serves two purposes for the
next iteration: (1) it is used to estimate if the *res* of a parameter should be increased
or decreased, i.e., step-size should be made smaller or larger and (2) to decide upon
the new *res* (Algorithm 1 Line 32). Certain conditioning on the SO value is required
to drive the feedback loop to minimize the difference and hit the coverage goal.
Therefore, it is passed onto the *Tuner* module.

Tuner

The *res* of a parameter for the next iteration is calculated in this block. First, the
Tuner checks if the *res* should be increased or decreased using Eq. 5.6.

$$direction = \begin{cases} increase & \text{if } SO > 0 \\ decrease & \text{if } SO < 0 \end{cases}. \tag{5.6}$$

If the SO is positive, the resolution is increased, i.e., the step-size is made smaller.
On the other hand, if it is negative, the resolution is decreased. Second, the amount of
increase or decrease in *res* is calculated. This is important as it affects the coverage
closure time. This can be done using different methods, but for our work, we only
consider increase or decrease of a fixed percentage. The advantage is that we do

---

[4] Please note that this does not mean that 100% coverage will always be achieved.

not require deep understanding of the DUV specification or complex mathematical models to achieve coverage closure.

Resolution Refinement

As mentioned earlier, in each iteration, the *res* is fine-tuned to generate new static stimuli parameters. The fine-tuning is done using a fixed percentage in our work, i.e., the amplitude *res* and frequency *res* are adjusted by 50% and 20% of the current value, respectively. This way, the resolution is systematically altered while slowly converging to 100% coverage. The new step-size for the parameters is set and the next iteration starts. All the input parameters are never adjusted simultaneously in any iteration; instead parameter S (Line 32) regulates which input parameter to adjust by switching to a new parameter every S iterations.

After N iterations, if the *hard coverage goals* are not achieved (Line 4), case C2 is considered, and the number of iterations N is increased iteratively (Line 35). However, if coverage goals are still not achieved, the simulation is terminated citing "recheck coverage goals" or "potential defect in DUV" (C3).

Capturing Unstable Behaviors

Algorithm 1 starts to analyze the unstable behaviors only if *coverage goals* are 100% (Lines 19–22 and Lines 26–29). It tries to achieve coverage closure for the *unstable coverage goals*, i.e., *overshoot/undershoot* and *secondary-extension* goals. The core idea is to identify early-on if the behavioral model shows unstable behavior at any given test input stimuli. It could occur because of the limitations of underlying algorithmic implementation, approximation of values using a certain equation system, or in some cases a certain configuration of the DUV. In this chapter, we classify the unstable behavior as *overshoot, undershoot,* and *undefined*. *Overshoot* refers to the DUV behavior if it goes above the maximum specified value. *Undershoot* refers to the DUV behavior if it goes below the minimum specified value. *Undefined* refers to the DUV behavior that is not defined at the selected *resolution*. The *unstable coverage goals* handle the *overshoot* and *undershoot* behavior, and *secondary-extension* handles the *undefined* behavior. The process is the same as presented earlier for *coverage goals*, and the only difference is that the number of iterations N is reduced to 10 (Line 18). The reasons are: (1) *unstable coverage goals* are also capturing DUV behavior during *coverage goals* closure, so some executions are already done and (2) the *unstable coverage goals* are derived to identify if the RF amplifier shows any unstable behavior. Please note that it is quite possible that none of the *unstable coverage goals* are achieved. In other words, we are defining here the goals that should not be hit in case of a correct DUV. In the next section, we illustrate the LCDG approach using our running example of LNA from Sect. 5.2.1 with gain G_2.

Illustration

We set N to 5, S to 1, and *COUNT* to 4. After the first iteration, we observe that the functional coverage for gain (G_2) in our running example is only 20% (Table 5.8). Out of five *output coverage bins*, only one is covered. Hence, according to Sect. 5.4.2, case C1, coverage analysis needs to fine-tune the stimuli parameter *resolution*. Out of five coverage holes, the first coverage hole, i.e., 16.0 dB, is selected as the coverage goal for the next iteration. The analysis searches for the nearest neighbor in DUV output trace (Table 5.11), i.e., least *Euclidean distance*. It comes out to be 16.6 dB. The nearest neighbor is subtracted from the coverage hole (Eq. 5.5) to calculate the difference. Since the difference is negative, the resolution should be decreased for iteration 2 (Eq. 5.6), i.e., step-size should be made bigger. Hence, *res* is adjusted for the amplitude by 50%, i.e., 1.5. New static stimuli parameters are generated using the refined *res* and the whole process repeats. The output coverage report for the second iteration is shown in Table 5.12. We do not observe any increase in the coverage; hence, the coverage goal remains the same and the whole process repeats. Iteration 3 shows no increase. In iteration 4 (Table 5.13), we observe an increase in coverage, i.e., 60%. Furthermore, we observe that not only was our coverage goal covered, but one other bin, i.e., 19.0 dB, has also been covered. In iteration 5, again the coverage does not change (not shown), and the analysis increases N to 10. In iteration 6 (Table 5.14), all the output coverage bins have been covered.

For *unstable coverage goals*, we observe the *overshoot/undershoot* and *secondary-extension* coverage reports, i.e., Tables 5.9 and 5.10, respectively. First, the analysis tries to achieve closure for *overshoot/undershoot* coverage goals. After

Table 5.12 LNA gain (G_2) output coverage report—iteration 2

gain_cvp "gain_values" iteration 1			
Description	Value	#Hits	Hit
16.0 dB	16.0	0	x
17.0 dB	17.0	0	x
18.0 dB	18.0	7	✓
19.0 dB	19.0	0	x
20.0 dB	20.0	0	x

20%

Resolution = 1.0 dB

Table 5.13 LNA gain (G_2) output coverage report—iteration 4

gain_cvp "gain_values" iteration 1			
Description	Value	#Hits	Hit
16.0 dB	16.0	5	✓
17.0 dB	17.0	0	x
18.0 dB	18.0	7	✓
19.0 dB	19.0	5	✓
20.0 dB	20.0	0	x

60%

Resolution = 1.0 dB

Table 5.14 LNA gain (G_2)
output coverage
report—iteration 6

		gain_cvp "gain_values" iteration 1			
		Description	Value	#Hits	Hit
100%		16.0 dB	16.0	5	✓
		17.0 dB	17.0	21	✓
		18.0 dB	18.0	7	✓
		19.0 dB	19.0	5	✓
		20.0 dB	20.0	16	✓

Resolution = 1.0 dB

10 more iterations, we observe that the *overshoot/undershoot* coverage has not increased. Furthermore, the *secondary-extension* goals are observed (Table 5.15), and we see a 100% coverage. Hence, no further refinement is required and the analysis stops. Please note that the complete LCDG approach is automated and requires no manual assistance.

A part of the cross-coverage report is displayed in Table 5.16 that shows the cross-coverage between checkers, input power, and gain. The first row of Table 5.16 can be interpreted as follows, the checker indicates the correct functionality of the DUV with *pass*, the input power at the time when the correct functionality was observed was 30 dBm, and the output gain was 17.0 dB. The functional coverage not only shows the exact behavior, but also the complete range of operation of the DUV, which is generally required. Furthermore, an excerpt of another cross-coverage report is shown in Table 5.17 that shows the cross-coverage between frequency, input power, and gain. One interesting thing that can be seen from Table 5.17 is that at the same input power of 40 dBm, two different coverage bins are covered. This is because the frequency of the signal was changed. This also shows the advantage of early verification, as exact input stimuli parameters can be observed that exercised a particular behavior. Furthermore, looking at the number of hits, we observe that many coverage goals are hit multiple times (as expected). This is because when resolution is refined, an overlap of values is expected between iterations.

5.4.3   *Experimental Evaluation*

In this section, we present the experiments to show the efficacy of our proposed LCDG approach for the early verification of *linear*, *non-linear*, and *unstable* behaviors of RF amplifiers. We consider seven configurable system-level models of LNA provided by our industrial collaboration partner. The next section provides details on the LNA models and their configurations. Then, we present the results obtained by using the proposed LCDG approach. We found a serious bug in the underlying implementation of the LNA using the proposed LCDG approach, which escaped the extensive verification performed by our industrial partner. Finally, we show the general quality of our proposed LCDG approach for system-level

Table 5.15 LNA gain (G$_2$)—secondary-extension goals

	gain_cvp "gain_values" iteration 1			
	Description	Value	#Hits	Hit
100%	16.25 dB	[16.25:16.49]	2	✓
	16.5 dB	[16.5:16.74]	6	✓
	16.75 dB	[16.75:16.99]	6	✓
	17.25 dB	[17.25:17.49]	1	✓
	17.5 dB	[17.5:17.74]	21	✓
	17.75 dB	[17.75:17.99]	3	✓
	18.25 dB	[18.25:18.49]	4	✓
	18.5 dB	[18.5:18.74]	4	✓
	18.75 dB	[18.75:18.99]	7	✓
	19.25 dB	[19.25:19.49]	7	✓
	19.5 dB	[19.5:19.74]	4	✓
	19.75 dB	[19.75:19.99]	4	✓

Resolution = 0.25 dB, *interval resolution* = 0.24 dB

Table 5.16 Cross-coverage (excerpt)—checker vs. input power vs. gain

	Checker vs. input power vs. gain		
	Description	#Hits	Hit
92.4%	Pass × 30 dBm × 17.0 dB	23	✓
	Pass × 37 dBm × 18.5 dB	21	✓
	Pass × 40 dBm × 19.0 dB	11	✓
	Pass × 40 dBm × 16.5 dB	21	✓

Table 5.17 Cross-coverage (excerpt)—frequency vs. input power vs. gain

	Frequency vs. input power vs. gain		
	Description	#Hits	Hit
92.4%	5 KHz × 30 dBm × 17.0 dB	23	✓
	11 KHz × 37 dBm × 18.5 dB	21	✓
	8 KHz × 40 dBm × 19.0 dB	11	✓
	14 KHz × 40 dBm × 16.5 dB	21	✓

verification. For this, we inject a fault in one of the LNA models and demonstrate that the LCDG approach was able to detect it.

LNA Models and Experimental Setup

In this section, we consider as DUVs seven system-level industrial LNA models. The behavioral models are implemented in SystemC AMS by our industry partner using the well-known concepts from [19, 81]. The simulations are carried out using the commercial tool COSIDE [115]. The specific configurations of the system-level models (denoted from *A* to *G*) adhere to the specifications given in Table 5.18. Columns 2–4 show gain (G) in dB, columns 5, 6 show 1 dB compression point, columns 7, 8 show IIP3 point in dBm, and columns 9, 10 show IIP2 point in dBm.

Table 5.18 Specifications of seven industrial LNAs

LNA	Gain			1 dB		IIP3		IIP2		Frequency		II (ohm)	OI (ohm)	Amplitude	
	min (dB)	typ Tip(dB)	max (dB)	min (dBm)	max (dBm)	min (dBm)	max (dBm)	min (dBm)	max (dBm)	min (MHz)	max (MHz)			min (v)	max (v)
A	17	18.2	20.5	29	34	38	43	42	47	0.005	0.02	50	50	0	5
B	17	19.8	25	20.1	24	29.7	33.6	33	38	0.05	0.1	50	50	0	3.3
C	14	16	17.2	17	20.3	26.4	30	29	34	0.003	0.013	50	50	0	5
D	11.5	13.2	14.2	33	38	42.6	47	45	49.7	5	20	50	50	0	5
E	20	22.1	25.6	30	33	40	42.6	41	44.5	400	500	50	50	0	3.3
F	13.1	15.0	17.8	27.8	33	37	42	40	45	3500	8000	50	50	0	3.3
G	15.8	18.2	20.1	15	19.8	24.6	29.2	29	34	0.005	0.02	50	50	0	1.2
FG	15.8	18.2	20.1	15	19.8	24.6	29.2	29	34	0.005	0.02	50	50	0	1.2

min minimum, *typ* typical, *max* maximum, *II/OI* input–output impedance *IIP2*: input second-order intercept point, *IIP3* input third-order intercept point

Columns 11, 12 show frequency in *Megahertz* (MHz), columns 13, 14 show input–output impedance in *ohms*, and the last two columns show allowed input signal amplitude range in volts (v). Please note that different LNA models are provided by the industry partner to check the effectiveness of the proposed ACDG approach. The total number of iterations N was set to 50, parameter S was set to 20% of N, and *COUNT* was set to 5. Please note that during execution only one specification is considered as a coverage goal at a time.

Case Study: Taylor Series Approximation

In this case study, we use as DUVs all the seven system-level models introduced in Table 5.18. The standard RF specifications of interest are gain, 1 dB compression point, and the intercept points—*input second/third-order intercept* (IIP2/3). The tables (Tables 5.19, 5.20, 5.21, and 5.22) show first 5 iterations of each specification's coverage progression. Column 1 shows the LNA models, columns 2,3 show amplitude resolution (*Ares*) and frequency resolution (*Fres*), respectively, and column 4 shows the total coverage (*cov*) achieved, gain coverage, 1 dB point coverage, IIP3 or IIP2 coverage, for one iteration. Columns 17 and 18 show the total iterations (TI) required to achieve coverage closure for *coverage goals* and the time taken (*coverage time* (CT)) to achieve those goals, respectively. The *unstable coverage goals*, i.e., *overshoot/undershoot coverage* (OUC), and *secondary-extension coverage* (SC) are shown in columns 19 and 20. Furthermore, column 21 shows the time required to achieve *unstable coverage goals*. The last column shows total simulation time (T) in seconds.

During gain coverage analysis (Tables 5.19 and 5.20), the iterations show the modification of amplitude only. This is because the frequency starts to change only after 10th iteration. It takes 13 iterations to achieve 100% coverage of gain for *LNA A* in 15 s. We observe that no *overshoot/undershoot coverage* goals are achieved. Furthermore, we see that all the *secondary-extension* goals are achieved, i.e., 100% coverage. It takes 8 s to check the *unstable coverage goals*, bringing the total simulation time to 23 s. *LNA C* is able to achieve 100% coverage in only 5 iterations and 3.5 s. The *overshoot/undershoot* coverage is 0% showing that *LNA C* is operating according to the specifications, and the *secondary-extension* coverage is only 75%.

Out of 7 LNA designs, the analysis was able to converge for 6 LNAs without doubling N. LNA F achieved the *coverage goals* in 137 iterations, with N increased to 200 and *COUNT* to 2. It was observed that because of a very wide range of frequencies for LNA F, i.e., 3.5 GHz to 8 GHz, the analysis was not able to converge in given iterations.

An interesting observation from Table 5.19 is the *LNA D* and its *overshoot/undershoot* goals. We observe a coverage of 31%. As motivated earlier, *overshoot/undershoot* coverage goals should not be hit during correct operation. Upon close manual inspection of the waveform (Fig. 5.9), we observed a slight overshoot of the gain curve for a short duration before settling to a stable value in the saturation

Table 5.19 LNAs case study—gain (G) progress over multiple iterations

| | Iterations | | | | | | | | | | | | | | | | | | | | | |
| | 1 | | | 2 | | | 3 | | | 4 | | | 5 | | | | | | | | |
LNA	Ares (v)	Fres (MHz)	cov (%)	Ares (v)	Fres (MHz)	cov (%)	Ares (v)	Fres (MHz)	cov (%)	Ares (v)	Fres (MHz)	cov (%)	Ares (v)	Fres (MHz)	cov (%)	TI	CT (s)	OUC	SC	UCT (s)	TT (s)
A	1	0.001	21	1.5	0.001	24	2.25	0.001	39	3.37	0.001	42	5	0.001	44	13	15	0	100	8	23
B	1	0.005	14	0.5	0.005	19	0.75	0.005	20	1.12	0.005	28	0.56	0.005	30	15	17.2	0	100	10.1	27.3
C	3	0.006	33	1.5	0.006	33	2.25	0.006	83	1.12	0.006	83	1.68	0.006	100	5	3.5	0	75	18.7	22.2
D	2.5	3	0	1.25	3	20	1.87	3	20	2.81	3	20	1.40	3	29	83	87	31	100	10	97
E	0.5	10	13	0.75	10	13	0.37	10	20	0.19	10	45	0.28	10	45	23	64.3	0	91.6	49.1	113.4
F	1	100	31	1.5	100	31	0.75	100	31	0.37	100	49	0.56	100	49	TO*	107.3	0	100	37.9	145
G	0.3	0.003	21	0.15	0.003	44	0.22	0.003	44	0.11	0.003	50	0.06	0.003	69	6	9.9	0	100	15.3	25.2
FG	0.3	0.003	21	0.15	0.003	44	0.22	0.003	44	0.11	0.003	50	0.06	0.003	69	6	9.9	0	100	15.3	25.2

Ares amplitude resolution, *cov* gain coverage, *Fres* frequency resolution, *TI* total iterations, *CT* coverage time, *UCT* unstable coverage time, *TT* total time, *TO* time out, *OUC* overshoot/undershoot coverage, *SC* secondary-extension coverage, *TO** time out, restart with $N = N \times 2$

Table 5.20 LNAs case study—1 dB compression point progress over multiple iterations

LNA	Iterations 1			2			3			4			5			TI	CT (s)	OUC	SC	UCT (s)	TT (s)
	Ares (v)	Fres (MHz)	cov (%)	Ares (v)	Fres (MHz)	cov (%)	Ares (v)	Fres (MHz)	cov (%)	Ares (v)	Fres (MHz)	cov (%)	Ares (v)	Fres (MHz)	cov (%)						
A	1	0.001	21	0.5	0.001	24	0.25	0.001	39	0.375	0.001	42	0.18	0.001	44	10	17.3	0	100	11.6	28.9
B	1	0.005	14	1.5	0.005	19	0.75	0.005	20	0.375	0.005	28	0.56	0.005	30	19	30.6	0	83.3	17.2	47.8
C	3	0.006	33	1.5	0.006	33	0.75	0.006	83	1.12	0.006	83	0.56	0.006	90	7	14.4	0	50	29.1	43.5
D	2.5	3	0	1.25	3	20	0.62	3	20	0.31	3	20	0.15	3	29	89	103.6	6.25	100	27.5	131.1
E	0.5	10	13	0.25	10	13	0.12	10	20	0.19	10	45	0.09	10	45	TO*	191.2	0	100	47.3	238.5
F	1	100	31	0.5	100	31	0.25	100	31	0.37	100	49	0.18	100	49	TO*	153.3	0	100	41.9	195.2
G	0.3	0.003	21	0.15	0.003	44	0.22	0.003	44	0.11	0.003	50	0.06	0.003	69	10	26.9	0	50	38.1	65
FG	0.3	0.003	0	0.15	0.003	0	0.22	0.003	0	0.11	0.003	0	0.06	0.003	0	TO	89.3	–	–	–	89.3

Ares amplitude resolution, *cov* gain coverage, *Fres* frequency resolution, *TI* total iterations, *CT* coverage time, *UCT* unstable coverage time, *TT* total time, *TO* time out, *OUC* overshoot/undershoot coverage, *TO** time out, restart with $N = N \times 2$, *SC* secondary-extension coverage

Table 5.21 LNAs case study—input third-order intercept point (IIP3) progress over multiple iterations

| LNA | Iterations | | | | | | | | | | | | | | | TI | CT (s) | OUC | SC | UCT (s) | TT (s) |
| | 1 | | | 2 | | | 3 | | | 4 | | | 5 | | | | | | | | |
	Ares (v)	Fres (MHz)	cov (%)	Ares (v)	Fres (MHz)	cov (%)	Ares (v)	Fres (MHz)	cov (%)	Ares (v)	Fres (MHz)	cov (%)	Ares (v)	Fres (MHz)	cov (%)						
A	1	0.001	10	1	0.0012	13	1	0.0014	22	1	0.0011	22	1	0.0009	22	37	43	0	58.3	32.8	75.8
B	1	0.005	0	1	0.006	0	1	0.0072	0	1	0.0057	0	1	0.0045	35	22	17	0	50	28.1	45.1
C	3	0.006	0	3	0.0048	19	3	0.0038	43	3	0.0046	43	3	0.0036	47	TO*	87	0	83.3	31.5	118.5
D	2.5	3	3	2.5	2.4	24	2.5	1.92	30	2.5	2.30	50	2.5	1.84	50	18	22	6.25	100	9.3	31.3
E	0.5	10	0	0.5	12	0	0.5	9.6	23	0.5	7.68	27	0.5	6.14	38	TO*	95.1	0	100	26.8	121.9
F	1	100	37	1	120	37	1	144	37	1	172.8	44	1	207.36	44	TO*	151	0	100	21.1	172.1
G	0.3	0.003	26	0.3	0.0024	50	0.3	0.0019	68	0.3	0.0015	68	0.3	0.0012	80	9	13	0	100	17.7	30.7
FG	0.3	0.003	26	0.3	0.0024	50	0.3	0.0019	50	0.3	0.0015	50	0.3	0.0018	50	TO	109	–	–	–	109

Ares amplitude resolution, cov gain coverage, Fres frequency resolution, TI total iterations, CT coverage time, UCT unstable coverage time, TT total time, TO time out, OUC overshoot/undershoot coverage, TO* time out, restart with $N = N \times 2$, SC secondary-extension coverage

Table 5.22 LNAs case study—input second-order intercept point (IIP2) progress over multiple iterations

LNA	1			2			3			4			5			TI	CT (s)	OUC	SC	UCT (s)	TT (s)
	Ares (v)	Fres (MHz)	cov (%)	Ares (v)	Fres (MHz)	cov (%)	Ares (v)	Fres (MHz)	cov (%)	Ares (v)	Fres (MHz)	cov (%)	Ares (v)	Fres (MHz)	cov (%)						
A	1	0.001	13	1	0.0012	19	1	0.0014	19	1	0.0011	32	1	0.0009	32	44	57.3	0	100	19.5	76.8
B	1	0.005	5	1	0.006	17	1	0.0072	17	1	0.0057	25	1	0.0045	29	17	29.7	0	100	11.9	41.6
C	3	0.006	9	3	0.0048	19	3	0.0038	19	3	0.0046	36	3	0.0036	39	TO*	101.4	0	66.6	45.7	147.1
D	2.5	3	10	2.5	2.4	10	2.5	1.92	31.5	2.5	2.30	39	2.5	1.84	43	27	37.8	18.7	0	39.1	76.9
E	0.5	10	0	0.5	12	9	0.5	9.6	15	0.5	7.68	24	0.5	6.14	24	TO*	103.9	0	75	41.5	145.4
F	1	100	15	1	120	30	1	144	30	1	172.8	39	1	207.36	59	TO*	131.8	0	41.6	34.4	166.2
G	0.3	0.003	19	0.3	0.0024	19	0.3	0.0019	35	0.3	0.0015	42	0.3	0.0012	76	14	17.6	0	83.3	27.1	44.7
FG	0.3	0.003	10	0.3	0.0024	10	0.3	0.0019	10	0.3	0.0015	43	0.3	0.0018	68	TO	97.4	–	–	–	97.4

Ares amplitude resolution, *cov* gain coverage, *Fres* frequency resolution, *TI* total iterations, *CT* coverage time, *UCT* unstable coverage time, *TT* total time, *TO* time out, *OUC* overshoot/undershoot coverage, *TO** time out, restart with $N = N \times 2$, *SC* secondary-extension coverage

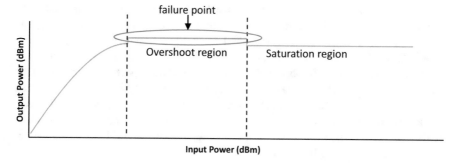

Fig. 5.9 Gain curve (output power vs. input power)—overshooting behavior due to Taylor series approximation

region. This faulty behavior is easy to miss and requires test-cases in a certain range of values and a certain configuration of DUV to exercise this behavior. For this reason, this behavior was not visible during the verification of other LNAs. During the verification, the amplitude of input test-stimuli was increased, which drove the amplifier in the non-linear region. In this region, the underlying approximation algorithm used in LNAs could not handle this non-linearity properly. As a consequence, we see an overshoot before the gain converges to a stable value. We observe that 31% of the *overshoot/undershoot* coverage goals were covered. Erroneously, the analog designer has chosen to use *Taylor Series Expansion* in the approximation algorithm of LNAs for the non-linearity region and did not consider the occurrence of higher-order polynomials [19]. This bug was discussed with our industrial cooperation partner, and they decided to fix the model accordingly using the concepts from [81]. We received the fixed system-level models. Now, with the fixed LNA models, no further bugs have been found with our proposed LCDG approach.

Furthermore, during IIP3 (Table 5.21) and IIP2 (Table 5.22) coverage analysis, the initial 5 iterations show the modification of frequency only. The amplitude changes after 10th iteration. Both the specifications show similar behavior in terms of coverage closure. Please note that the IIP2 and IIP3 points are measured using multi-tone signal source. Out of 7 LNA designs, the analysis was able to converge for 4 LNAs without doubling N. LNA C, LNA E, and LNA F required doubling of N iterations to achieve 100% coverage.

In the next section, we analyze the general quality of our approach in detecting bugs.

Case Study: Fault-Injection and LCDG

To show the general quality of our proposed LCDG approach to detect faults, we injected a fault in *LNA G*. The faulty LNA is called *LNA FG* (Table 5.19—last row). The nature of fault is similar to the one found in Sect. 5.3.1, i.e., the fault transformed the LNA FG into a linear device; hence, it never saturated. Interestingly, the fault injected into the LNA *FG* did not show any faulty behavior w.r.t. the gain; hence, we see 100% coverage. During the 1 dB compression point coverage analysis (Table 5.20), the LNA *FG* did not achieve 100% coverage even with an increased number of iterations. We observed from the coverage reports that the coverage stayed at 0%. On a closer look, we observed that the faulty behavior of the LNA *FG* is not letting the device saturate; hence, 1 dB compression points will never be achieved. The gain of the faulty LNA *FG* (actual gain) and bug-free LNA *G* (expected gain) turned out to be similar to Fig. 5.5, where the LNA *FG* does not saturate with increasing input power.

In summary, the proposed LCDG approach for RF amplifiers systematically and automatically increased the coverage to verify the *linear*, *non-linear*, and *unstable* behaviors of LNA models.

5.5   Summary

In this chapter, two enhanced functional coverage-driven methodologies are proposed for a thorough verification of RF amplifiers at system level. As a representative of RF amplifiers, system-level model of an industrial LNA was selected.

First, we proposed the AMS functional coverage-driven verification approach that elevated the main concepts of digital functional coverage to the SystemC AMS abstraction. Two coverage refinement parameters, *resolution* and *interval resolution*, were introduced to define the stimuli parameters and functional coverage bins at input and output. An AMS coverage analysis was proposed, which crosses input and output functional coverage and checkers, to systematically modify the refinement parameters to fully capture the specifications of the DUV. We showed the effectiveness and applicability of our approach on an industrial RF transmitter and receiver model.

Second, this chapter proposed a lightweight CDG approach to automatically verify the *linear*, *non-linear*, and *unstable* behaviors of RF amplifiers at system level. The approach leveraged functional coverage data in combination with the much simpler *Euclidean Distance* to achieve coverage closure. The LCDG approach put particular focus on capturing the unstable behavior of LNAs during the verification phase. During the extensive experiments, the LCDG approach was able to find a serious bug in the LNA implementation. The encouraging results show that LCDG can be effectively used to verify the *linear*, *non-linear*, and *unstable* behaviors of the RF amplifiers early-on in the design phase.

Chapter 6
Digital Early Security Validation

This chapter discusses novel security validation approaches to ensure high quality secure VPs using static and dynamic information flow analyses. As motivated earlier in Sect. 1.1, the increasing functionality and connectivity of embedded devices such as in the Internet-of-Things have raised their requirements on security significantly. This is due to the increasing amount of sensitive information and personal data being stored in those devices as well as the security-critical functions they perform. Secure embedded systems cannot be achieved by focusing on just the software or the hardware part, but rather require holistic approaches. As the number of vulnerabilities in software/firmware historically dominates, security strategies and validation approaches for embedded software have received considerably more attention and also benefited from traditional software security research. Formally verified operating systems (e.g., seL4) and compilers (e.g., CompCert) or mature information flow tracking approaches are just to name a few. However, hardware security is at least equally important since a potential hardware backdoor that allows unprivileged software access to confidential data will render all software/OS-level protection mechanisms useless. Consider for example a recent discovered exploitable bug [108] in the Actel ProASIC3 FPGA, which is claimed to be one of the most secure devices in the industry and being used in military and other critical application domains. A backdoor via JTAG allows an attacker to get hold of cryptography keys as well as other data from the FPGA.

Furthermore, fixing hardware security issues after deployment is always associated with very high cost. The reason is that the underlying SOC cannot be patched after production. Therefore, it is very critical for the embedded industry to move rapidly from considering security as an afterthought to integrating it in the SOC design process. Still, current practice and research on SOC security validation focuses mostly on the *Register Transfer Level* (RTL) and below. Many approaches for Trojan/backdoor identification and IC counterfeit detection exist. Verification of hardware security architectures (e.g., ARM TrustZone) and security modules starts only with the availability of RTL designs. While these steps are indispensable, ample

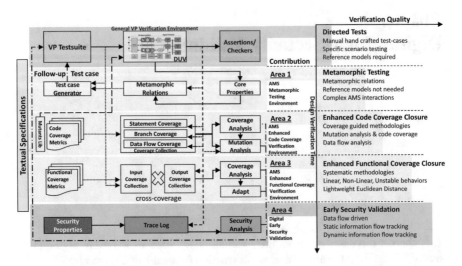

Fig. 6.1 Early security validation

opportunities to save time and cost by prioritizing security earlier at the system level have not been yet fully exploited. Hence, VP-based security validation would enable to detect and correct many SOC hardware security issues very early. In this regard, this chapter extends the general VP verification environment with several components in pink area as shown in Fig. 6.1. The security validation environment uses security properties, VP execution trace logs, and a combination of novel static and dynamic security analyses to thoroughly secure the VP. As a result, a wide variety of security vulnerabilities are covered.

In Sect. 6.1, we present a novel approach to SOC security validation at the system level using VPs. At the heart of the approach is a scalable static information flow analysis that can detect potential security breaches such as data leakage and untrusted access; confidentiality and integrity issues, respectively. We demonstrate the applicability of the approach on real-world VPs. This approach has been published in [61].

Then, in Sect. 6.2, we extend the static approach of Sect. 6.1 to a dynamic approach. We present the first dynamic information flow analysis at ESL. Our approach allows to validate the run-time behavior of a given SOC implemented using VPs against security threat models, such as confidentiality and integrity. Experiments show the applicability and efficacy of the proposed method on various VPs including a real-world system. This approach has been published in [40].

6.1   Static Information Flow Analysis

Information Flow Tracking (IFT) has become one of the key techniques in security research. The basic idea is to control how (labeled) information is propagated by the system under consideration. This tracking allows to enforce policies for secure information flow such as confidentiality and integrity. In this section, we present a novel *VP-based IFT* approach, which is the *the first of its kind*. Essentially, our approach operates directly on the SystemC VP models by combining several passes of static analysis. The main difficulties to be overcome here are to deal with the challenging language C++, which SystemC is based on, as well as the specific semantics of TLM on-chip communication via *TLM-2.0 payload*. To this end, we build on the flexible compiler infrastructure provided by LLVM/Clang to perform in interleaved manner connectivity analysis, access control extraction, call-graph analysis, data flow analysis, and static taint tracking to identify static paths that violate specified secure information flow properties. These potential vulnerable paths are reported back to user for further inspection. Our static analysis is *sound*, i.e., it never misses a violating path if such exists.

6.1.1   Approach Overview

In this section, we give a high-level overview of our approach starting with the threat model we consider. Then, we discuss a motivating example and show the overall workflow of the approach.

Threat Model

Considering a SOC, we want to protect assets like for instance: cryptographic keys, digital certificates, signatures, classified text, authentication data from id sensors, or register settings. Transporting the sensitive data is performed through buses between the different IP components of the SOC. Hence, for our proposed VP-based IFT approach, the analysis of the information flow using TLM communication between the IP blocks is of particular interest. In this context, confidentiality (an IP creates an unwanted information flow from a target IP in retrieving secret data which this IP is not allowed to access) and integrity (an IP presents itself as a different IP to create an information flow to some target IP to modify some data) are major security concerns.

Motivating Example

We present here a SystemC TLM-2.0 example (Listing 6.1) that will be used to showcase the main ideas of our approach throughout this section. The SystemC TLM-2.0 constructs and semantics necessary to understand the example will be explained as needed. The example presents a simplified SOC consisting of a regular MPU (Microprocessor) (MPU1), a trusted MPU (MPU2), a regular memory (Memory0), and a confidential memory with security keys (Memory1) as shown in Fig. 6.2. The buffer (Memory2) is initially not available. The modules are connected to an interconnect that routes transactions where the MPUs act as initiator and the memories as target. The communication uses a 32-bit address mode as follows: (1) bits 0–7—local address inside a memory; (2) bits 8–15—memory address; (3) bits 16–23—MPU ID; (4) bits 24–31—unused. Their behavior is implemented in thread functions (MPU1: `thread\_proc()` Line 3—MPU2: `thread\_proc()` Line 16), and *b_transport* functions (Interconnect: `b\_transport()` Line 26—Memory: `b\_transport()` Line 46).

The MPUs execute instructions that initiate TLM-2.0 transactions (i.e., read or write) to the memory. In this illustrative example, the actual instruction behavior is abstracted away as we only focus on the communication. The Interconnect receives the transaction, checks the address generated by MPUs (Line 27, to Line 35), and accordingly routes the TLM 2.0 transaction to the corresponding memory. The memory receives the transaction, checks the *cmd* from transaction, and writes to (Line 56) or reads from (Line 54) the memory.

A more formal representation of the information flow policies of the SOC will be given later in Sect. 6.1.2. The intuition here is that MPU1 should not be able to access Memory1 that holds the security keys. The access control policies that are implemented in the Interconnect to enforce this are as follows:

1. If ((mpu_nr == 1) && (mem_nr == 0))
2. If ((mpu_nr == 2) && (mem_nr == 1))

Such access control policies are commonly implemented in components with routing functions. Intuitively, the access control policies, when properly implemented, would be enough to achieve the isolation of the untrusted MPU1. This will be confirmed later by our analysis as detailed in Sect. 6.1.2. Now consider a new

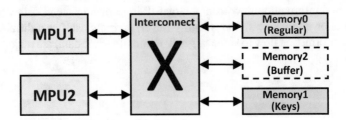

Fig. 6.2 Motivating example

```
1   struct MPU1: sc_module {                34    (*initiator_socket[1])−>b_transport(trans,delay);
2   //...                                   35    else
3   void thread_proc() {                    36    trans.set_response_status(
4   tlm::tlm_generic_payload* trans = new    37          tlm::TLM_OK_RESPONSE );
        tlm::tlm_generic_payload;           38  }
5   //...                                   39  sc_dt::uint64 addrs;
6   trans−>set_data_ptr(reinterpret_cast<unsigned  40  sc_dt::uint64 masked_address;
        char*>(&data));                     41  unsigned int mem_nr;
7   //...                                   42  int mpu_nr;
8   socket−>b_transport(*trans, delay);     43  };
9   //...
10  }                                       44  struct Memory : sc_module {
11  int data;                               45  //...
12  };                                      46  virtual void b_transport
13                                                  (tlm::tlm_generic_payload& trans,
14  struct MPU2 : sc_module {                       sc_time& delay) {
15  //...                                    47  tlm::tlm_command cmd =
16  void thread_proc(){                              trans.get_command();
17  //...                                    48  sc_dt::uint64 adr = trans.get_address();
18  }                                        49  unsigned char* ptr = trans.get_data_ptr();
19  //...                                    50  unsigned int len = trans.get_data_length();
20  };                                       51  unsigned char* byt =
21                                                  trans.get_byte_enable_ptr();
22  struct Interconnect : sc_module {        52  unsigned int wid =
23  SC_HAS_PROCESS(Interconnect);                    trans.get_streaming_width();
24  Interconnect(sc_module_name name) { ... }  53  //...
25  //...                                    54  if ( cmd == tlm::TLM_READ_COMMAND )
26  void b_transport(int id,                 55    memcpy(ptr, &mem[adr], len);
        tlm::tlm_generic_payload& trans,    56  else if ( cmd ==
        sc_time& delay) {                            tlm::TLM_WRITE_COMMAND )
27  masked_addrs = trans.get_address();      57    memcpy(&mem[adr], ptr, len);
28  mpu_nr = (masked_addrs >>16) & 0xFF;     58  //...
29  address = masked_addrs & 0xFF;           59  trans.set_response_status(
30  mem_nr = (masked_addrs >> 8) & 0x3;              tlm::TLM_OK_RESPONSE );
31  if ( (mpu_nr == 1) && (mem_nr == 0) )    60  }
32  (*initiator_socket[0])−>b_transport(trans,delay);  61  int mem[MEMORY_SIZE];
33  else if ( (mpu_nr == 2) && (mem_nr == 1) )  62  };
```

Listing 6.1 SystemC TLM 2.0 example with two MPUs and two memories

scenario that a designer decides to add an additional buffer (Memory2), shared by
both MPUs, for buffering data for the overall performance enhancement. However,
with the shared buffer, indirect information flow between Memory1 and MPU1
exists. A software adversary can use MPU2 to read the confidential keys from
Memory1 and write them into the buffer. Then MPU1 can read the keys from
the buffer. Such indirect information flow (via another IP) is not trivial to detect,
especially without an automated analysis like our proposed VP-based IFT approach.

Overall Workflow

The overall workflow of our data flow driven information flow approach for
SystemC TLM 2.0 is shown in Fig. 6.3. The approach is based on a static analysis;
hence, it needs to be only run once to validate the information flow security
properties. Essentially, our approach uses data flow analysis to perform static taint

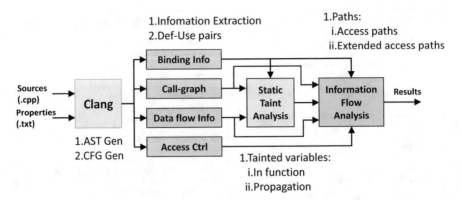

Fig. 6.3 Information flow analysis overview

analysis and combines this information with system binding information and call-graphs to validate the security properties defined for a SOC. First the *security properties* are read, followed by identification of source and sink (destination) of each property to be verified in different paths. From the elaboration phase of the system, connectivity of VPs is identified (*binding information*). This helps in identifying how data flows through the system. It is followed by *access control information* extraction. It identifies the set of all access control policies implemented inside a system. In the next step, construction of call-graphs is performed to be used in analysis at the end. The *data flow analysis* identifies the set of all data flow values computed at different points in a system. Due to static nature, the analysis computes an over-approximation of the data flow values. This data flow information is extended to perform taint analysis between source and sink of each property. All the tainted variables from the source are identified w.r.t. SystemC.

In the next step, all the information from aforementioned steps is evaluated and combined to obtain the final access paths and extended paths between source and sink for each property. Both the paths are then filtered using predicates to ensure the validation of property. Essentially, the result shows which security properties have been satisfied and which are not satisfied at the end. A property is satisfied iff one of the three following conditions is satisfied: (1) There exists no access paths or extended access paths between a source and sink. (2) There exists no access paths or extended access paths between any tainted variable from source and sink. (3) The predicates of the property fail completely.

Our approach identifies the failing paths for each unsatisfied security property and allows the verification engineer to focus his efforts to improve the design by improving either access control policies or information flow policies.

In the following, we detail the ingredients of our approach as well as demonstrate them using the motivating example.

6.1.2 Data Flow Driven Information Flow Analysis

Information Extraction

Information Flow Property Specification

Information flow can be used to model various security properties i.e., confidentiality, integrity, availability, isolation, and hardware trojans, etc. Our approach focuses only on two widely used security properties: confidentiality and integrity based on the principle of non-interference. An information flow property defines a data flow relation among two entities, *source* and *sink*, where the data is allowed or disallowed to flow under some certain conditions. These properties need to be defined in a way to capture the complete flow without missing out on any information. For SystemC TLM 2.0, the classical property specification does not work as the classical specification always considers an input port and an output port of the system as untrusted/trusted information sources; whereas in SystemC TLM 2.0, the VPs interact through transactions; hence, the property specifications and definitions need to adapt. Therefore, we define an information flow property as a tuple *(source, sourcep, op, sink, sinkp)* that, depending on the property being defined (confidentiality/integrity), comprises of the following elements:

1. *source*—The generation point of a transaction/variable inside a VP
2. *sourcep*—A predicate associated with *source*, which specifies under what condition is the data valid at *source*
3. *op*—Operator such as *no flow*
4. *sink*—The point in program where the *source* assigns some value
5. *sinkp*—A predicate associated with *sink*, which specifies under what condition is the data valid at *sink*

Information is said to flow between *source* and *sink* when there is a static path between the two entities, and *sourcep* predicate and *sinkp* predicate are satisfied, whereas information is said not to flow when there is no static path even when *sourcep* and *sinkp* are satisfied. This could be due to several underlying implementation issues.

For our analysis, we only focus on *op no flow*, which inhibits the flow of information between *source* and *sink* to make the VP secure. Also, other *op* such as *flow* can be defined as a dual of *no flow* for every case.

VP Binding Information

In a SOC, VPs are connected to each other in a certain way that affects how information is propagated. Our analysis defines a connection of two VPs when their corresponding sockets (*initiator_socket* and *target_socket*) are connected using SystemC TLM 2.0 *b_transport* function call. The way these functions are registered and bound during elaboration phase is vital because each VP in TLM

2.0 setting contains a *b_transport* function. If binding information is not available, the connectivity of modules cannot be identified statically (before execution). Also, without this information, a *b_transport* function call from a wrong VP can be analyzed because of similar function name, resulting in further over-approximation. This binding information helps in constructing call-graph.

Access Control Extraction

Our analysis extracts all the access control conditions from each function of a VP. An access control condition is defined as a condition in *if-else* control flow structure, or in *while* and *for* loops. Our analysis extracts this information without expanding any functions and stores it. This is useful for the analysis in that it tells which VPs are allowed to access which VPs in a complete SOC. Access control conditions can be based on VP id, VP address, or the socket id, etc. Generally, this information is embedded in interconnects that act as transaction routers in SystemC TLM 2.0.

Call-Graph Construction

Our static analysis performs call-graph construction once in the start. Call-sites are not expanded at this stage, i.e., before data flow analysis. Also, there are a lot of SystemC specific function calls that are added to the call-graph, but they are never expanded. For the commonly occurring system function calls, their behavior is already defined inside the analysis, e.g., *memcpy()*. This call-graph is used to guide the analysis to propagate the information to the correct function. Hence, it uses binding information to correctly identify the function call from the correct VP.

Static Analysis

Data Flow Analysis

A data flow analysis algorithm takes as input the SOC under test to compute test objectives for each VP (i.e., def–use pairs). A *reaching uses* procedure—an instance of data flow analysis techniques—is used to identify test objectives (i.e., def–use pairs) for a VP, which actually answers such a question: for each variable defined, which uses can potentially use the values? Our data flow analysis is inspired from Sect. 3.3, but we do not use the associations as defined. Rather, we only define definition–use (*def–use*) pairs according to their classification to help in our analysis, like the *def–use* pairs across threads. The analysis identifies all possible *def–use* pairs of a VP by performing intra-function analysis and inter-thread analysis. Inter-function analysis is deliberately not performed as it will be compensated for during the taint analysis.

Static Taint Analysis

Static taint analysis identifies how a single VP or a variable inside a VP affects or taints other VPs inside a system. The core idea behind the taint analysis w.r.t. SystemC TLM 2.0 is that any VP inside the SOC when generates a transaction, all the entities, and paths it takes to reach the destination entity are also compromised. Because if the generated transaction is malicious, the malicious value can propagate throughout the system and at the end leak some data. Not only does the transaction leak the data, but all the variables in between also become potential security risks. Static taint analysis may be viewed as a conservative approximation of the full verification of non-interference or the more general concept of secure information flow. Because information flow in a system cannot be verified by examining a single execution trace of that system, the results of taint analysis will necessarily reflect approximate information regarding the information flow characteristics of the SOC to which it is applied. Our taint analysis uses the data flow information (*def–use*) pairs and combines it with security properties information.

The *source* of each property acts as a taint source, it can be a transaction generated in a VP, or it can be an internal variable or register. The *sink* in each property acts as the final destination where taint source propagates. We define a tainted variable as a variable that is affected by taint source directly or indirectly, for, e.g., a direct assignment from *source* or an assignment from a variable that was tainted by *source*. Hence, for a taint to propagate to *sink*, it is not necessary that the taint source itself has to propagate, rather any of its tainted variables can also propagate. Our static analysis builds a graph of all tainted variables belonging to one taint source. The transfer function for assignments taints the left hand side if any of the operands on the right hand side is tainted. Assignments to arrays and memories are treated conservatively by tainting the entire array/memory because at static time we do not know the exact address location that will be accessed.

Information Flow Analysis

Our information flow analysis interleaves call-graph information, access control information, VP binding information, security properties, and data flow analysis to detect vulnerable paths. We define two kinds of paths: (1) Access paths—A path between *source* and *sink*; (2) Extended access paths—A combination of access paths.

Each access path is defined as a combination of nodes where each node represents an entity. A node can be from one of the four categories: (1) Taint source node—*tsrc*; (2) Function call node—*fcall*; (3) Access control node—*actrl*; (4) Taint sink node—*tsink*.

A taint source node always defines a starting point of an access path, and taint sink node always ends the access path. If there is no taint sink found (i.e., the information does not reach it), the access path is not created at all. A function call node is created when taint propagates to the next function, and an access control

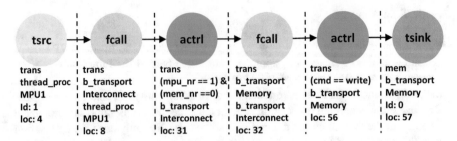

Fig. 6.4 One access path for example shown in Listing 6.1

node is created when the path encounters access control condition. Each node carries with itself some useful information that helps in the analysis. For example, the node contains the taint source name, the calling function's name, the called function's name, their corresponding classes, and ids etc. (see Fig. 6.4).

The analysis is done in three stages: (1) Forward analysis; (2) Backward analysis; (3) Predication. The forward analysis is carried out first, starting from first function where the taint source is defined. By definition of a taint source, we mean when the *source* is created (formal parameters of a function definition do not qualify for that). A taint source node is added as the starting point of the path. Then data flow analysis information, i.e., def–use pairs information, is used to deduce if taint sink lies in the same function. If it does, a taint sink node is added to the path; hence, an access path is created between these two points. If the sink is not in the same function, the call-graph information is combined with data flow information to find the taint propagation. In case of a function call, a function call node is added to the path, and the actual parameters are mapped onto the formal parameters of the function followed by retrieving the data flow information of the callee. If an access control exists, a corresponding node is inserted in the path. Once the complete access paths for all the security properties are created, the backward analysis starts. During the creation of an access path, all the irrelevant information is abstracted away, i.e., if a function is being called without taint source/variables, or if there exists an access control not affecting taint source/variable, because such information does not help in any way in information flow analysis.

The purpose of backward analysis is to detect if there is any other access path leading to the same taint sink. Our analysis uses forward analysis information and overlaps the access paths to check which access path ends at the sink. The starting point of the path can be any point, except the taint source, because in that case a loop will be created. When the path is found, it is added to the path found by forward analysis. Then it is checked if this new path propagates this information anywhere else. If it propagates, the paths are added to the initial path, and at the end, a complete extended access path is created. One thing to keep in mind, extended paths will always be equal to or greater than access paths, but never less because essentially extended paths are a combination of access paths, and if there is only one combination possible for each path, then access paths equal extended paths.

In the last stage, predication is performed by applying the *sourcep* predicate and *destp* predicate for each property onto the extended paths found. A property is satisfied if the predicates are true and there is no static path between *source* and *sink*, or if the predicates are false.

Illustration

We use the same SystemC TLM 2.0 example as shown in Listing 6.1 to illustrate our methodology. The information flow properties are deduced from the policies defined in Sect. 6.1.1, i.e.:

1. source: trans, sourcep: mpu_nr == 1, op: no flow, sink: mem, sinkp: mem_nr = 1
2. source: trans, sourcep: mpu_nr == 2, op: no flow, sink: mem, sinkp: mem_nr = 0

When the information flow analysis is executed, the first property is picked, and all the paths with *trans* as *source* and *mem* as *sink* are searched. The availability of paths signifies that there is a static path between a *source* and a *sink*, it can be within one function or spread over multiple functions, and it can be passing through various access control mechanisms. The analysis identifies four paths originating from MPU1 and four paths originating from MPU2. One of the paths originating from MPU1 and terminating at Memory0 is shown in Fig. 6.4.

The starting node, i.e., *tsrc*, contains the information that it represents variable *trans*, and its originating function is *thread_proc*, which belongs to class MPU1 with registered id 1 (id extracted from elaboration phase). Similarly, other nodes also contain useful information.

In the next step, the analysis picks first path and checks if the *sink* of first path overlaps with the *sink* of any other path, and finds that MPU2 accesses *mem*. So, this new path is appended to the actual path. In the third step, the analysis checks if the newly added path accesses *sink*, *mem* at any other location in the system, and finds that MPU2 accesses *mem* in Memory1. It recursively checks for any other similar paths and does not find anything. The analysis finds all the extended paths in similar fashion and stores them. The same is done for second property. We get the following extended paths for first property:

1. MPU1 $\longrightarrow$ Memory0 $\longrightarrow$ MPU2 $\longrightarrow$ Memory1
2. MPU1 $\longrightarrow$ Memory1 $\longrightarrow$ MPU2 $\longrightarrow$ Memory0
3. MPU2 $\longrightarrow$ Memory0 $\longrightarrow$ MPU1 $\longrightarrow$ Memory1
4. MPU2 $\longrightarrow$ Memory1 $\longrightarrow$ MPU1 $\longrightarrow$ Memory0

The long arrows signify flow of information between two VPs. This information flow could be passing through access control policies, or it could be a direct path. For example, for the first extended path, *trans* generates in MPU1, and it propagates to *mem* in Memory0. Then this *mem* in Memory0 can be accessed by MPU2's *trans*. Finally, *trans* in MPU2 can propagate to *mem* in Memory1. Essentially, each VP's name shown in extended path is an access path as shown in Fig. 6.4. If more VPs are connected in between *source* and *sink*, this extended path would also get long.

At the end, predication is performed by applying source predicates *sourcep* and sink predicates *sinkp*. This process narrows down the number of paths as the predicates are applied. For example, initially first extended path related to first property is analyzed. The *sourcep* predicate states that the *trans* is from MPU1; hence using the information gathered in aforementioned sections, it is deduced that indeed this predicate is true. Because *sourcep* has been satisfied, the next predicate *sinkp* is checked, which states that *mem* should be in Memory1. The analysis does not find *mem* in MPU1. It proceeds further to Memory0 while passing through the access control policies. It checks if MPU1 is allowed to access Memory0 (Line 31— Listing 6.1), and finds that it is allowed. Then, the analysis deduces that Memory0 is not allowed to be accessed by MPU2 (Line 33—Listing 6.1), and returns false. When the access control condition is false, that means the information will not be allowed to flow further, and hence, we stop searching in this path further. The second path is analyzed, and again it is found that MPU1 is not allowed to access MEMORY1. Hence, there exists no path that allows MPU1 to access Memory1. In the similar fashion, property 2 is also satisfied. Hence, the the system is shown to be secure.

Now consider a new scenario that a designer decides to add an additional buffer (Memory2), shared by both MPUs, for buffering data for the overall performance enhancement. The *b_transport()* function of interconnect (Line 26—Listing 6.1) is updated with the following information as shown in Listing 6.2. The rest of the function implementation remains the same.

```
1    ....
2    if ( (mpu_nr == 1) && (mem_nr == 0) )
3        (*initiator_socket[0])->b_transport(trans,delay);
4    else if ( (mpu_nr == 2) && (mem_nr == 1) )
5        (*initiator_socket[1])->b_transport(trans,delay);
6    else if ( mem_nr == 2 )  (*@\label{line:line_cond3}@*)
7        (*initiator_socket[2])->b_transport(trans,delay);
8    else
9        trans.set_response_status( tlm::TLM_OK_RESPONSE );
10   ....
```

Listing 6.2 SystemC TLM 2.0 example with two MPUs and three memories

When our analysis is executed once again, it detects that both the security properties fail now. Upon inspection of the extended paths, it is found that now MPU1 can access Memory1 using the shared *buffer—Memory2*. Similarly, MPU2 can access Memory0. This indirect information flow (via another IP) is not trivial to detect. To solve the problem, for example, another buffer should be added, and the sharing of resources is prohibited using the *b_transport* code shown in Listing 6.3 where each memory is given exclusive access. A more general solution is to add a memory management unit.

```
1    ....
2    if ( (mpu_nr == 1) && (mem_nr == 0) )
3        (*initiator_socket[0])->b_transport(trans,delay);
4    else if ( (mpu_nr == 2) && (mem_nr == 1) )
5        (*initiator_socket[1])->b_transport(trans,delay);
6    else if ( (mpu_nr == 1) && (mem_nr == 2) )
7        (*initiator_socket[2])->b_transport(trans,delay);
8    else if ( (mpu_nr == 2) && (mem_nr == 3) )
9        (*initiator_socket[3])->b_transport(trans,delay);
10   else
11       trans.set_response_status( tlm::TLM_OK_RESPONSE );
12   ....
```

Listing 6.3   SystemC TLM 2.0 example with four memories

The information flow properties are also updated with two new properties:

1. source: trans, sourcep: mpu_nr == 1, op: no flow, sink: mem, sinkp: mem_nr = 3
2. source: trans, sourcep: mpu_nr == 2, op: no flow, sink: mem, sinkp: mem_nr = 2

After executing the analysis again, all the properties are found to be satisfied again. This example demonstrates the capability of our approach to detect *omission of access control policies*.

Implementation Details

In this section, we describe the implementation details of our methodology briefly. The static information flow analysis is implemented using the LibTooling library for Clang compiler [83]. Clang generates an Abstract Syntax Tree (AST) of the SystemC TLM 2.0 source code. The AST is parsed to extract the required information to perform static analysis. The implementation parses the AST multiple times and in each iteration extracts a different set of information. We next discuss important implementation details.

In the first iteration, the VP binding information is extracted by looking for the AST nodes related to elaboration phase. The binding of VPs is possible in different ways depending on the implementation of the SystemC design, i.e., using *bind* keyword or using AMBA binders.

In the next iterations, access control information and call-graph information are extracted using the aforementioned data extraction methods. The locations of *use* are also extracted to help with the analysis, especially the bounds of access control information that define a context of program statements. All the extracted information is stored in data structures for performing data flow analysis, taint analysis, and information flow analysis.

6.1.3   Experimental Results

In this section, we present a case study to demonstrate our information flow analysis approach for SystemC. We consider the LEON3-based VP SOCRocket [100] that has been modeled in SystemC TLM 2.0. The complete VP consists of more than 50, 000 lines of code. The VP combines several IPs working together in master or slave mode. The VP is designed to be *bus-centric*, i.e., the IP cores are connected through an on-chip bus. The AMBA-2.0 AHB/APB (Advanced High-performance Bus/Advanced Peripheral Bus) bus is used as the common on-chip bus. In the VP, the LEON3 processor is directly connected with AHB as *AHBMaster* device, and AHB/APB Bridge is connected as a *AHBSlave* controller that controls the communication between AMBA AHB and AMBA APB devices. Various TLM IPs are connected to the AMBA APB bus such as UART, GPTimer, and IRQMP. A memory controller is connected as *AHBSlave* with AHB bus that serves several memories. Essentially, AMBA-2.0 AHB/APB buses are complex interconnects that take in the transactions generated by the connected IPs and forward them to the intended IP based on the address and/or sockets. In the following, we report results for three experiments integrating different IPs into SOCRocket.

CRYPTO AES IP

In the first case study, we consider the case of integrating two TLM IPs with SOCRocket: (1) A TLM IP *CRYPTO-AES*, a cryptographic hardware accelerator implementing AES-128 algorithm designed specifically to compute *cipher text* for the given *plain text* efficiently; (2) A secure memory IP *SEC-MEMORY* pre-loaded with cipher keys.

The *CRYPTO-AES* engine works in *Cipher Block Chaining (CBC)* mode, i.e., an *initialization vector* and a *plaint text* are given as input to the IP, where *plain text* and *initialization vector* are XORed before being written on input of IP, *key* is read from *SEC-MEMORY*, and the IP generates a *cipher text*. This *cipher text* is fed back to the cryptographic engine as the new *initialization vector* for the next iteration. The IP *CRYPTO-AES* computes its round keys on the fly, instead of computing them beforehand and storing them in *SEC-MEMORY*. The LEON3 processor initializes *CRYPTO-AES* by configuring its configuration registers.

Because of the nature of *SEC-MEMORY*, i.e., it stores cryptographic keys, our information flow policy is that the LEON3 processor should not be allowed to read (confidentiality) or write (integrity) these keys from *SEC-MEMORY*. Hence, the following security property covering both confidentiality and integrity is derived:

1. source: trans, sourcep: IP_ADDR == 0x1, op: no flow, sink: memory_buffer, sinkp: MEMORY_ADDR == 0x4

IP_ADDR = 0x1 refers to the address of LEON3, and *MEMORY_ADDR = 0x4* to the address of *SEC-MEMORY*. With our approach, we observed that the security property failed. Our methodology was able to report information flow between LEON3 and *SEC-MEMORY* through the *debug* interface. Normally, the *debug* interface is constrained to output limited information or dummy information in case of cryptographic algorithms, but it was not the case. We fixed the failing path by restricting the *debug* interface access to *CRYPTO-AES* only. The property was satisfied on next analysis run. Sometimes, these intentional (supposedly non-malicious) flaws can occur in hardware designs, especially when the design team includes undocumented features for assistance in testing. These flaws can be exploited to get access to trusted data. The analysis reported a computation of 37 access paths and 79 extended access paths. In these paths, only two extended paths (one for read access, and one for write access to debug interface) allowed the information flow, while the others did not. The analysis took 51.63 s to report the results. 0% false positives were reported because of concrete path conditions.

Near-Field Communication IP

For the second case study, we integrate a *near-field communication* (NFC) interface IP with SOCRocket. Essentially, it is a communication protocol that enables two devices to communicate when brought in close proximity. NFC is now widely used in smartphones as a mode of contactless payment system, for sharing files and photos between two devices, and as e-ids (electronic ids), etc. Because of high-speed communication, the data is stored in memory using *Direct Memory Access* (DMA). Most current SOCs involve DMA to the memory through a dedicated DMA controller to reduce the workload on the processor cores.

Due to the nature of NFC IP and the overall system security, the security policy states that system-specific addresses (could be pointing to configuration registers of SOC) in memory should not be accessed by the NFC IP (read or write). Therefore, we derive the following security property:

1. source: trans, sourcep: IP_ADDR == 0x5, op: no flow, sink: main_memory, sinkp: MEM_SPACE_ADDR < 0x000F0000

IP_ADDR = 0x5 refers to NFC IP in SOCRocket, whereas *MEM_SPACE_ADDR < 0x000F0000* classifies the system-specific addresses the NFC IP shall not access.

Our approach detects an access path between NFC and *main_memory* via DMA; thus, the security property is violated. The reason is that the requested DMA address (from the NFC IP) is not checked against disallowed address ranges (missing bounds). The reported numbers of access paths and extended access paths were 37 and 80, respectively. The analysis took 55.01 s to invalidate the property. 0% false positives were reported because of concrete path conditions. It clearly shows that IP development team and SOC integration team should collaborate actively to avoid such security vulnerabilities.

Smart Card Reader IP

For the third case study, we integrate a smart card reader into the system. It reads the data from card and stores it in a secure section of on-chip memory (access control implemented by address range specification). Three security policies are specified: (1) The on-chip processor (LEON3) should not be allowed to read (confidentiality)/write (integrity) this secure portion of memory. (2) Smart card reader should not read (confidentiality) or write (integrity) SRAM. (3) Smart card reader should not read (confidentiality) or write (integrity) *Read Only Memory* (ROM) (contains LEON3 configuration, etc.).

After running our analysis, we observed that the first and second security properties are satisfied because of the strict access control policy implemented in the Memory Controller (*MCtrl*) in SOCRocket. But the third property fails. The *MCtrl* disallows regular ROM accesses by any IP other than LEON3, but this constraint is not present on DMA. This violation of the security property, reported by our analysis, could potentially be exploited by a hardware trojan to get DMA access for ROM. The reported numbers of access paths and extended access paths were 37 and 83, respectively. The analysis took 74.35 s to validate all three properties. Due to the use of concrete path conditions, 0% false positives were reported.

6.2 Dynamic Information Flow Analysis

In this section, we propose a VP-based IFT approach that operates directly on the binary VP models, i.e., third-party VP models without source code are analyzed using a combination of dynamic IFT and post-execution analyses. This does not come without challenges, i.e., SystemC specific language constructs, and the TLM-2.0 semantics (e.g., *interoperability layer*) and timing models. We take advantage of the *GNU Debugger* (GDB) as a virtual execution platform to trace TLM transactions, followed by transaction extraction, security property generation, and finally VP model validation against security properties, to identify the TLM transactions, and the dynamic paths they take to violate the security properties. The focus of the proposed methodology is to detect the security violations related to the most occurring threat models: confidentiality and integrity. The precise violation paths are reported back to the verification engineer to either replace the (third party) VP model or update the security policy.

6.2.1 Motivating Example and Threat Models

In this section, we introduce the threat models considered for dynamic IFT and show a motivating example to demonstrate how the threat models affect security of a SOC.

Threat Models

To show the importance and criticality of security validation in a given SOC design, it is necessary to identify: (1) which types of assets must be protected and (2) what kind of threats we are protecting against. Assets, such as sensitive configuration registers, cryptography keys, digital signatures, etc., are the critical and confidential information that must be protected against unauthorized access. According to [34, 61], the threat models of leaking information in a given SOC design at ESL can be divided into two main categories:

- **Confidentiality:** Data of secure IP (e.g., data in a secure memory) is retrieved by an unauthorized IP.
- **Integrity:** Data of secure IP is modified by an unauthorized IP.

For a given SOC design, two common potential sources of the aforementioned threat models are: (1) The design contains no (strong) security policies (i.e., access control or information flow policy). This makes the design exploitable by vulnerabilities. For example, when a SOC includes a hardware IP purchased from an untrusted third-party vendor, the IP can contain malicious part to leak the confidential data. In the same way, an incorrect initialization (either by an adversary involved in the IP design process or unintentionally) of the SOC firmware (e.g., memory configuration file) can cause an unauthorized IP to access the sensitive data in the secure memory. In both cases, a well-implemented security policy in the SOC could prevent the sensitive data from being leaked. (2) The existing SOC is extended/modified, but its security policy is not updated. Especially in case that the design team decides to improve the existing SOC by adding new IPs (e.g., an accelerator, a processor, or a memory), the previous security policies may not be sufficient to protect the sensitive data against leakage. Therefore, for a given SOC, the security validation analysis must be performed to ensure that the secure assets cannot be inferred either with *direct access* (direct communication of two IPs) or through *indirect access* (IP collusion).

```
1   void foo_64 (fstream &init_key){
2       /* ... */
3       unsigned char key1[8]  = init1;
4       unsigned char key2[8]  = init2;
5       unsigned char final_key[8] = {0};
6       /* ... */
7       final_key = key_gen (key1,key2);
8       /* ... */
9       return ;
10  }
11  /*The function is upgraded to generate more robust key*/
12  void foo_256 (fstream &init_key){
13      /* ... */
14      unsigned char key1[32] = init1;
15      unsigned char key2[32] = init2;
16      unsigned char key3[32] = init3;
17      unsigned char final_key[32];
```

```
18      /* ... */
19      final_key = key_gen (key1,key2,key3);
20      /* ... */
21      return;
22   }
```

Listing 6.4 A part of *RISC-CPU* running software of the motivating example.

Motivating Example

Consider the *RISC-32 SOC* model shown in Fig. 6.5. The hardware part consists of two initiator modules: *RISC-CPU* and *Initiator_0*, a generic loosely timed interconnect *LT_BUS* and two memories *M0_regular* (regular memory) and *M2_secure* (secure memory). The interconnect module is an executable binary [7] (i.e., the source code is not available) accompanied with a specifications sheet. It can support up to four initiators and eight target modules based on the following configuration: (1) the target modules connected to the *LT_BUS* initiator ports 1 and 8 are only accessible by initiator modules connected to the *LT_BUS* target ports 1 and 4, respectively, and (2) the target modules connected to the *LT_BUS* initiator ports 2 to 7 are shared with all four initiators. *M2_secure* stores sensitive data (connected to initiator port 1 of the *LT_BUS*), and *M0_regular* (connected to initiator port 8 of the *LT_BUS*) is used to store normal computation results. The *RISC-CPU* module (connected to target port 1 of the *LT_BUS*) is a standard commercial 32-bit RISC-CPU provided by [2] and extended to support TLM-2.0 and C++ as its running software. The initiator *Initiator_0* is a processing element that is connected to target

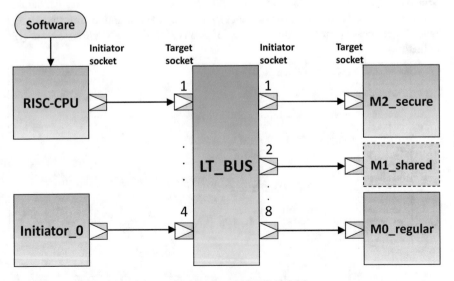

Fig. 6.5 The architecture of the motivation example *RISC-32 SOC*

port 4 of the *LT_BUS*. Memory *M1_shared* (connected to initiator port 2 of the *LT_BUS*) is not initially available.

The software part (running on the *RISC-CPU*) includes a key generation routine for an encryption algorithm (to simplify, the actual functionality is abstracted away), a compiler (to translate the software into *RISC-CPU* instructions), and a memory configuration file (used by the compiler for memory allocation). The memory configuration file for the *RISC-CPU* initially includes the information that the only available memory is *M2_secure*. To reduce the risk of run-time errors due to memory management, memory is allocated at compile time that is mostly used in SOC designs. Listing 6.4 demonstrates a part of the running software *foo_64* (Lines 1–10) of the *RISC-CPU* generating a 64-bit encryption key. Variables *key*1 (Line 3) and *key*2 (Line 4) are initialized by *init*1 and *init*2 that are extracted from the *init_key* (Line 1) file. The final key *final_key* (Line 7) is generated after some intermediate computation by *key_gen* function. As *M2_secure* is the only available memory for the *RISC-CPU* in the memory configuration file, all variables are mapped by the compiler into this memory.

According to the aforementioned configuration w.r.t. the interconnect specification, the following security policies are satisfied for the *RISC-32 SOC* design:

- *M0_regular* is only accessible by *initiator_0*.
- *M2_secure* is only accessible by *RISC-CPU*.

Now consider the scenario that the software is upgraded (*foo_256*—Line 12 in Listing 6.4) to generate a more robust key by increasing its length (from 64 bits to 256 bits) and modifying the *key_gen* (Line 19 in Listing 6.4) function by adding one more initial key *key*3 to its input arguments. After compiling the new software, an error is generated by the compiler stating that the memory is not sufficient. To overcome this problem and to increase the overall performance of the design, memory *M1_shared* is added to the *RISC-32 SOC*. The memory configuration file needs to be updated, allowing the compiler to use the new memory space to map data. Thus, it might be possible that some variables (such as *final_key*—Line 19 in Listing 6.4) containing the secure data (that have been already located in *M2_secure* before adding the new memory) are now mapped into the *M1_shared*. Therefore, at run-time, an unwanted information flow can cause sensitive data to be read from the secure memory and then written to the shared memory. The main reason for such vulnerability is that the interconnect security policy is not updated after inserting a new IP (in the example at hand *M1_shared*). Thus, the underlying hardware can be used as a gateway by an unprivileged software to leak the confidential data. Since the security policy of the *LT_BUS* has not been updated, the running software (i.e., application) can use the *RISC-CPU* to read the sensitive data from *M2_secure* and write it to *M1_shared*. Subsequently, *initiator_0* can read the secure data from *M1_shared*.

Detecting this type of indirect access even for a simple SOC model is not a trivial task as it cannot be detected either by functional verification methods (as the functionality of the SOC model is still correct) or using static security validation analysis. In case that the address of transactions is defined at run-time, e.g., generated either explicitly by initiator modules (based on some dynamic computation) or implicitly by its running software (like in the example above), static analysis approaches are not able to detect this security violation. Hence, a dynamic information flow analysis is required such that the run-time behavior of a given SOC is verified against the threat models in both *direct access* and *indirect access* cases.

6.2.2   Dynamic IFT Methodology

Overall Workflow

Figure 6.6 provides an overview of the proposed methodology. As can be seen, the methodology consists of four main phases:

1. Tracking transactions of a given VP model at run-time
2. Translating the transactions into a set of access paths
3. Generating security properties from the design's security rules
4. Validating the translated model of the design against the security properties

The information related to each transaction is extracted only once during the execution time and translated into a set of access paths. Each path identifies an information flow from the *source* (the creation point of a transaction by an initiator IP) and *destination* (the final point of the transaction that is in a target IP). The design security rules are translated into a set of security properties based on the direct and indirect access scenarios. In the final phase, the security validation analysis ensures that there is no path where an unauthorized *source* is connected to a protected *destination*. The methodology identifies the failing paths for each unsatisfied security property, allowing the design team to focus on the exact vulnerable points of the design and improve it either by enhancing access control policies or information flow policies. In the following, each phase of the proposed method is explained in detail and illustrated using the motivating example of Sect. 6.2.1.

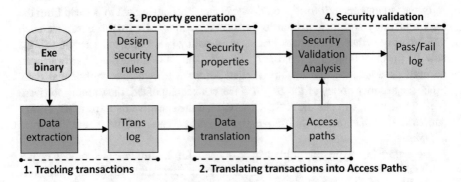

Fig. 6.6 The architecture of the proposed methodology

Tracking Transactions

IPs in a SOC at VP level communicate through TLM transactions. To validate that
no information leaks, a complete trace of all transactions is required. Especially,
the *indirect access* can only be detected when the whole behavior of the design
including all intermediate communications of IPs is analyzed. To do this, we take
advantage of utilizing the GDB as a virtual execution platform to access the run-time
information of a given TLM design.

As illustrated in Fig. 6.7, first, the debug symbols of the design are analyzed
to extract the static information. This refers to the modules' name, corresponding

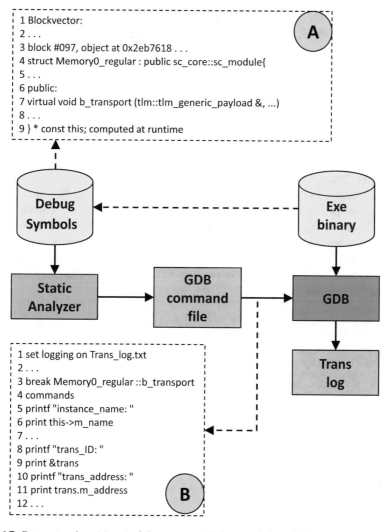

Fig. 6.7 Data extraction. (**a**) part of the generated *Debug symbols* and (**b**) part of the generated
GDB command file of the *RISC-32 SOC* model

member functions and transaction object, and its attributes. The static information is used to program the debugger in order to trace transactions at run-time. Hence, a combination of the extracted static information is used with the basic commands of GDB (e.g., setting breakpoints for a function based on its name, printing the name, type, and value of its variables) to automatically generate a *GDB command file* (GDB script). For example, to extract all transactions related to the *RISC-32 SOC*'s target module Memory0_regular, all functions of the module in which a transaction object is referenced need to be traced. This is performed by finding blocks that contain the information of module's function from the design's debug symbols (e.g., Fig. 6.7a). The function's input arguments and local variables are also in the block. Based on the extracted data, the b_transport function of the module must be traced as well as the transaction object trans defined as its input argument. To trace the function, a breakpoint is set at the beginning of the function body (Fig. 6.7b, Line 3). For this breakpoint, a set of commands (Lines 4–11, Fig. 6.7b) is defined that is executed whenever the breakpoint is triggered at run-time. The information that needs to be extracted can be defined within these commands and contains information such as the instance name of the module (Lines 5 and 6, Fig. 6.7b) or the transaction's reference address, and address attribute (Lines 8–11, Fig. 6.7b).

Second, the execution is controlled by the pre-programmed GDB to automatically extract the run-time information of the design and store it in a log file. Our information extraction process is inspired from [41], but we do not extract the whole transaction lifetime (i.e., the information from time of creation till end) as defined. Rather, the GDB is programmed to only retrieve and log the dynamic information related to the *source* and *destination* of a transaction. Moreover, for each transaction, we only extract the attributes, e.g., address and *data_length*, and its related parameters such as reference address of transaction object, transition phase, and return value of function call that are required for our security validation analysis. This speeds up our information flow analysis as the amount of data that needs to be extracted is reduced.

```
1    AP_RISC32-SOC = {
2    P1=(0x6761B0_0 -> initiator_0::init_0::
3    thread_process -> READ -> M0_regular::mem0::
4    b_transport-> 0x004 -> 4-> 0ns),
5
6    P2=(0x683C10_0 -> RISC-CPU::init_1::
7    thread_process -> READ -> M2_secure::mem2::
8    b_transport-> 0x010 -> 4-> 100ns),
9
10   P3=(0x6761B0_1 -> initiator_0::initi_0::
11   thread_process -> WRITE -> M0_regular::mem0::
12   b_transport-> 0x008 -> 4-> 200ns),
13
14   P4=(0x683C10_1 -> RISC-CPU::init_1::
15   thread_process -> WRITE -> M1_shared::mem1::
16   b_transport-> 0x004 -> 4-> 300ns),
17
18   P5=(0x6761B0_2 -> initiator_0::init_0::
19   thread_process -> READ -> M1_shared::mem1::
20   b_transport-> 0x004 -> 4-> 400ns)}
```

Listing 6.5 A part of generated access path of the *RISC-32-SOC* design.

Translating Transactions into Access Paths

The SOC behavior can be seen in terms of abstract communication paths illustrating how transactions flow through the system (from *source* to *destination*). As the security rules of a given design are defined based on communication between the initiator and the target modules, each access path must be defined based on the information related to the *source* and *destination* of the corresponding transaction.

The behavior of a given TLM design is defined based on a set of *access paths* (AP) using the following definition.

Definition 1

$$AP = \{P_i \mid P_i = \{TID \rightarrow IM \rightarrow cmd \rightarrow TM \rightarrow$$
$$adrs \rightarrow L \rightarrow ST\} ; \ 1 \leq i \leq n_T\}, \tag{6.1}$$

where:

- *TID* is the transaction ID including transaction reference address (to distinguish the generated transactions of different initiator modules).
- *IM/TM* is initiator/target module that consists of its root name, the instance name, and the name of its function call (which defines the start/end point of the transaction).
- *cmd* is the command attribute of the transaction.
- *adrs* is the address attribute of the transaction (indicating transaction address in *TM*).
- *L* is the *data_length* attribute of the transaction.
- *ST* is the simulation time stamp of transactions.
- n_T is the number of transactions.

The log file is analyzed to extract all access paths embedded in the transactions' lifetime and present them based on Definition 1. Each member of the *AP* set in (6.1) indicates a direct flow of information between *IM* and *TM*, thus declared as a direct path. A combination of three direct paths specifies an indirect information flow if it enables an initiator to access the data of a target that is inaccessible directly. For instance, Listing 6.5 illustrates a part of the generated *RISC-32-SOC* design's access paths. In this set, *P*1, *P*2, … and *P*5 present five direct paths. Paths *P*1, *P*3, and *P*5 indicate three transactions generated by *initiator_0* to access data of regular and shared memories *M0_regular* and *M1_shared*, respectively. For indirect path, consider the combination of paths *P*2, *P*4, and *P*5 creating an indirect data flow between *initiator_0* and *M2_secure*. It demonstrates that in path *P*2 the *RISC_CPU* generates a transaction to read data from secure memory *M2_secure*. In path *P*4, *RISC_CPU* writes the read secure data into shared memory *M1_shared*, and finally in path *P*5, the secure data is read by *initiator_0*. Thus, *initiator_0* could access the secure data using *RISC_CPU*.

Algorithm 2: Direct security property generation

Data: targets' lists of initiator modules TL_IM
Result: Security properties to validate direct access paths DSP
1 $DTL \leftarrow$ extracting all target modules of design form TL_IM;
2 $i \leftarrow 0$;
3 $DSP \leftarrow \emptyset$;
4 **foreach** *target list TL of initiator module IM in TL_IM* **do**
5     $FTL_{IM} \leftarrow DTL - TL$;
6   **foreach** *target T in FTL_{IM}* **do**
7         $p_i \leftarrow (IM, T, \{R/W\})$;
8         $add(DSP, p_i)$;
9         $i \leftarrow i + 1$;

Algorithm 3: Indirect security property generation

Data: Secure target list STL and targets' lists of initiator modules TL_IM
Result: Security properties to validate indirect acess paths ISP
1 $DTL \leftarrow$ extracting all target modules of design form TL_IM;
2 $i \leftarrow 0$;
3 $ISP \leftarrow \emptyset$;
4 **foreach** *target module T in STL* **do**
5   **foreach** *initiator module IM in TL_IM* **do**
6   **if** *T in target list TL of IM* **then**
7             $seq_1 \leftarrow (IM, T, \{R\})$;
8   **foreach** *target module T' in $DTL - STL$* **do**
9   **if** *T' in target list TL of IM* **then**
10                     $seq_2 \leftarrow (IM, T', \{W\})$;
11   **foreach** *initiator module IM' in $TL\_IM - IM$* **do**
12   **if** *T' in target list TL of IM'* **then**
13                             $seq_3 \leftarrow (IM', T', \{R\})$;
14                             $p_i \leftarrow (seq_1, seq_2, seq_3)$;
15                             $add(ISP, p_i)$;
16                             $i \leftarrow i + 1$;

Security Property Generation

The design security rules need to be expressed in a well-defined format in order to validate the SOC. The security rules are usually written in a textbook specification (defined as reference model) where designers use it to implement the access control policies of the interconnect module or other mechanisms to prevent information leakage. The security rules of the SOC are given as input. This includes two elements: (1) *Target List of Initiator Module* (TL_IM). For each initiator, it includes a list of target modules allowed to access. (2) *Secure Target List* (STL) including the target modules containing critical data. The formal definition of design specification in the *design security rules* file is as follows:

Definition 2

$$STL = \{t_i \mid t_i \text{ is a secure memory} ; \ 1 \leq i \leq n\}$$

$$TL\_IM = \{IM_i \mid IM_i \rightarrow \{t_j ; \ 1 \leq j \leq m\} ; \ 1 \leq i \leq n_I\} \qquad (6.2)$$

where n and m illustrate the number of secure target modules and the number of targets that initiator module IM_i is allowed to access, respectively. n_I indicates the number of initiator modules of the design.

Algorithm 2 illustrates the algorithm of generating *Direct Security Property* (DSP) from the *design security rules*. First, the *TL_IM* list is analyzed to extract all target modules of the design and store them in the *Design Target List* (DTL) (Line 1). For each initiator module *IM* of the design, a *Forbidden Target List* (FTL_{IM}) is generated containing the target modules that the initiator is not allowed to access (Lines 4 and 5). This is performed by eliminating the *Target List* (TL) of initiator *IM* (defined in *TL_IM*) from *DTL*. Finally, for each target in *FTL*, a property is generated including the initiator name, target name, and the access mode (Lines 6–9). Each direct security property p_i in DSP identifies that there must be no flow (e.g., read or write access) from initiator *IM* to target *T* (Lines 7 and 8).

The generation of the *Indirect Security Property* (ISP) is explained in Algorithm 3. Similar to Algorithm 1, it starts with extracting the design's target modules from the *TL_IM* set (Line 1). In order to detect the indirect scenario, the generated ISP must include three sequences as follows:

- The first sequence (seq_1) indicates a read access from a secure initiator (i.e., the initiator module allowed to access a secure memory) to its secure target (Lines 4–7).
- The second sequence (seq_2) identifies a write access from the secure initiator of seq_1 to a non-secure memory (Lines 8 and 9).
- The third sequence (seq_3) is a read access from a non-secure initiator (i.e., the initiator module that is not allowed to access a secure memory) to the target of seq_2 (Lines 10 and 11).

Consider the running example *RISC-32-SOC* (Fig. 6.5). The *design security rules* file is defined as follows:

$$STL = \{M2\_secure\}$$

$$TL\_IM = \{Initiator\_0 \rightarrow \{M0\_regular, M1\_shared\},$$

$$RISC\_CPU \rightarrow \{M1\_shared, M2\_secure\}\} \qquad (6.3)$$

Based on its *design security rules*, the *DSP* and *ISP* are generated as follows:

$$DSP = \{p_1, p_2 \mid p_1 \rightarrow (Initiator\_0, M2\_secure, \{R/W\}),$$

$$p_2 \rightarrow (RISC\_CPU, M0\_regular, \{R/W\})\}$$

$$ISP = \{p_1 \mid p_1 \rightarrow ((RISC\_CPU, M2\_secure, \{R\}),$$
$$(RISC\_CPU, M1\_shared, \{W\}),$$
$$(Initiator\_0, M1\_shared, \{R\}))\} \tag{6.4}$$

Note that when a part of memory (range of memory addresses) is defined as the secure part, the *design security rules* can be modified to support it.

Algorithm 4: Security violation detection

Data: set of access paths AP and security properties DSP and ISP
Result: violated access paths V_{AP}
/* validating AP against direct security properties   */
1 **foreach** *direct property dp in DSP* **do**
2   **foreach** *path p in AP* **do**
3   **if** *(dp in p)* **then**
4             $add(V_{AP}, p)$;

/* validating AP against indirect security properties   */
5 **foreach** *indirect property ip in ISP* **do**
6   **foreach** *path p in AP* **do**
7   **if** *(ip.seq$_1$ in p)* **then**
8   **foreach** *path p' in AP* **do**
9   **if** *(ip.seq$_2$ in p') & (p.ST > p'.ST)* **then**
10                     $add(V_{AP}.suspected, (p, p'))$;
11   **foreach** *path p'' in AP* **do**
12   **if** *(ip.seq$_3$ in p'') & (p'.ST > p''.ST) & (ip.seq$_2$.adrs == ip.seq$_3$.adrs)* **then**
13                         $add(V_{AP}, (p, p', p''))$;

Security Validation

In case of *direct access*, for each property in *DSP*, the *AP* set is traversed (Algorithm 4—Lines 1–4) in order to find a security policy violation. The detection of a property violation in this case signifies that there is a direct path between a non-secure initiator module and a secure memory. All violated paths are stored in *Violated Access Path* (V_{AP}) set to be reported to designers. To detect a property violation related to *indirect access* (Lines 5–13), the *AP* set of design is analyzed to find a combination of direct access paths in which the whole or a part of secure data is leaked. We split the security violation in this case into two cases that are *suspected* and *violated*.

The former refers to the two sequential accesses of an initiator where it reads a sensitive data from a secure memory and then writes the data to a non-secure memory. We define these two accesses as a suspect case as it may lead to information leakage. The *suspected* case is detected when the first two sequences (seq_1 and seq_2) of an indirect security property are violated (Lines 7–10). The paths related to this violation are stored in $V_{AP}.suspected$ set. When a non-secure initiator reads the sensitive data from the non-secure memory written by the secure initiator (Lines 11–13), the *suspected* case becomes a real indirect security violation that is defined as *violated* case. Therefore, the analysis reports the paths of violated properties in V_{AP} set, e.g., verifying the *AP* set of the running example *RISC-32-SOC* (Listing 6.5) against its security properties shows that an ISP is violated. The access paths reported by the analysis illustrate that the secure data is leaked by a combination of direct paths $\{P2, P4, P5\}$.

To know how much of the secret information is leaked (considered as leakage depth analysis), the data length attribute (L) of the transaction specifies the size of data that is read from or written to the memory block by the initiator module. In case that the violated property is a DSP, parameters $adrs$ and L (in the related path) signify the depth of leakage. Regarding *indirect access*, the analysis must be performed on more than one path as the violated ISP includes three sequences. The access paths of the last two sequences seq_2 and seq_3 contain the information of writing the secure data to and reading it from a non-secure memory, respectively. The $adrs$ parameter of both sequences identifies the accessed block of memory by the initiators. If data length in seq_3 (L_3) is equal to or smaller than data length in seq_2 (L_2), the depth of leakage is equal to L_3 otherwise is L_2.

6.2.3 Experimental Evaluation

The experimental evaluation of the proposed approach has been performed on various VP models implemented in SystemC TLM-2.0. The experiments cover both the generality and scalability of the proposed method. The former refers to the security validation of designs [7] implementing various aspects of the TLM-2.0 standard (core interfaces, the base protocol, and coding styles). The latter refers to the security validation of a real-world VP-based SOC [100].

To evaluate the quality of the proposed method, we consider two possible security scenarios covering both confidentiality and integrity threat models:

- S1: Modifying a SOC without updating its security policies (similar to the motivation example scenario in Sect. 6.2.1)
- S2: Incorrect initialization or update of the SOC firmware (i.e., memory config-uration file)

We apply the aforementioned security scenarios to VP models and verify their security policies against their security rules. All the experiments are carried out on a PC equipped with 8 GB RAM and an Intel core i7-2760QM CPU running at 2.4 GHz.

Case Studies

The experimental results for different types of SOC benchmarks are shown in Table 6.1. The first two columns list the names and lines of code for each VP model, respectively. Note that the original model of each SOC is modified by integrating different initiator and target modules with its interconnect to support various security scenarios. Column *TM* presents the timing model of each design. Column *SSc* shows the security scenario applied to each VP model. Column *#IP* lists the number of IPs for each VP model followed by the total number of IPs, secure initiator, and secure memory in *Total*, *SI*, and *SM*, respectively. The column *#Trans* shows the number of extracted transactions for each design. Columns *#ISP* and *#DSP* present the result of verifying each VP model against the indirect and direct security properties, respectively. For both columns, *Total*, *Pass*, and *Fail* illustrate the number of generated, satisfied, and violated properties, respectively. The execution time of the proposed method is reported in column *Time* including the data extraction *ET*, validation process *VT*, and total execution time *Total* of the method. The time reported in *VT* covers phases 2–4 of the proposed method. Column *CET* shows the total time of each design's compilation and execution without any instrumentation.

For a real-world experiment, we modified the LEON3-based VP SOCRocket [100] by integrating three TLM-2.0 IPs with its AMBA-2.0 AHB (Advanced High-performance Bus): (1) a synthesizable SytemC IP *AES_core* [99] used as a hardware accelerator to implement AES-128 encryption algorithm, (2) two secure memories *Mem_Secure1* and *Mem_Secure2* initialized by *cryptography keys* and *plain texts*, respectively. The VP itself is implemented in SystemC TLM-2.0 consisting of more than 50,000 lines of code. It includes several IPs working together in master (e.g., initiator modules *LEON3* processor, *ahbin1* and *ahbin2*) or slave (e.g., target modules *AHBMem1* and *AHBMem2*) mode connecting to the on-chip bus AMBA-2.0. The communication uses a 32-bit address mode where the 12 most significant bits are used to specify the memory address. For brevity, we refrain from giving a detailed introduction to the whole VP. The necessary parts to understand the experiment are explained as needed.

The expected security policies of the VP are as follows:

- *Mem_Secure1* and *Mem_Secure2* are secure memories and only accessible by *AES_core*.
- *AHBMem1* and *AHBMem2* are regular memories and only accessible by *LEON3*, *ahbin1* and *ahbin2*.

Initially, memory *AHBMem2* is not available, and memory configuration is defined based on the aforementioned security policies. At the beginning of execution, initiator modules read the memory configuration file to extract the range of memory addresses that they are allowed to access. The *AES_core* module executes the standard AES-128 encryption algorithm using the *initialized keys* and *plain texts* stored in secure memories *Mem_Secure1* and *Mem_Secure2*, respectively.

Table 6.1 Experimental results for all case studies

VP model[a]	LoC	TM	SSc	#IP			#Trans	#ISP			#DSP			Time (m:s)			CET (s)
				Total	SI	SM		Total	Pass	Fail	Total	Pass	Fail	ET	VT	Total	
Routing-model[b]	656	LT	S1	9	2	2	150	24	20	4	5	4	1	0:41	0:15	0:56	1.9
RISC32-SOC[b,c]	3150	LT	S1	6	1	1	110	1	0	1	2	2	0	0:35	0:4	0:39	2.7
AES128-SOC[b]	4742	AT	S1,S2	10	2	3	350	33	31	2	13	8	5	3:28	0:34	4:02	16.2
RISC32-SOC[b,c]	4850	AT	S1	19	3	3	733	324	304	20	36	34	2	9:16	0:52	10.08	23.8
Locking-two[b]	5830	LT/AT	S2	13	2	2	450	48	39	9	14	10	4	6:42	0:29	7:11	22.2
Locking-auto[b]	6959	LT/AT	S2	15	2	3	370	96	85	11	20	18	2	8:33	0:39	9:12	22.7
SOCRocket[d,e]	>50000	LT/AT	S2	21	1	3	1100	180	168	12	30	28	2	25:56	1:44	27:40	43.1

[a]The VP models are provided by [b][7], [c][2], [d][100] [e][99] and modified using a different combination of initiator and target modules to support various security scenarios

LoC lines of code, *TM* timing model, *SSc* security scenario, *#IP* the number of intellectual properties, *SI* secure initiator, *SM* secure memory, *#Trans* the number of transactions, *ISP* indirect security property, *DSP* direct security property, *ET* extraction time of the proposed method (phase 1), *VT* validation time of the proposed method (phases 2–4), *CET* compilation and execution time of each VP model using GCC

In order to increase the overall performance of the system, consider the scenario that the design team decides to integrate *AHBMem2* with the *AT-bus* (AHB). In order to make the new memory accessible by other initiators, the memory configuration file needs to be updated. The expected update from the design team for memory configuration is as follows: in *AHBMem2*, memory blocks:

- (0xA0000000 to 0xA0000BB4) are shared among *LEON3*, *ahbin1*, and *ahbin2*.
- (0xA0000BB5 to 0xA0000DE6) are only accessible by *AES_core*.
- (0xA0000DE7 to 0xA0000FFF) are shared between *ahbin1* and *ahbin2*.

The security scenario is that the memory configuration file is incorrectly updated (either by malicious insider on the design team or unintentionally) where in *AHBMem2*, memory blocks:

- (0xA0000000 to 0xA0000BB4) are shared among *LEON3*, *ahbin1*, and *ahbin2*.
- (**0xA0000BA5** to 0xA0000DE6) only accessible by *AES_core*.
- (0xA0000DE7 to 0xA0000FFF) are shared between *ahbin1* and *ahbin2*.

The wrong memory configuration in this case is difficult to be detected by conventional functional tests because of two reasons: First, this fault does not affect the functionality of the system as the location of storing variables is only shifted in the same memory. Second, from the functional point of view, *AHBMem2* is reachable by all expected initiators, and the transactions data is still stored in the expected memory (*AHBMem2*). From the security point of view, the secure memories *Mem_Secure1, Mem_Secure2* and *AHBMem2* (0xA0000BB5 to 0xA0000DE6) must be protected against unauthorized access. Hence, the *Design security rules* of the *SOCRocket* are defined as shown in (6.5):

$$ST L = \{Mem\_Secure1, Mem\_Secure2, AHBMem2\ (0xA0000BB5-$$

$$0xA0000DE6)\}$$

$$T L\_IM = \{AES\_core \rightarrow \{Mem\_Secure1, Mem\_Secure2,$$

$$AHBMem2\ (0xA0000BB5 - 0xA0000DE6)\},$$

$$LEON \rightarrow \{AHBMem1, AHBMem2\ (0xA0000000 - 0xA0000BB4)\},$$

$$ahbin1 \rightarrow \{AHBMem1, AHBMem2\ (0xA0000000 - 0xA0000BB4;$$

$$0xA0000DE7 - 0xA0000FFF)\},$$

$$ahbin2 \rightarrow \{AHBMem1, AHBMem2\ (0xA0000000 - 0xA0000BB4;$$

$$0xA0000DE7 - 0xA0000FFF)\}\} \tag{6.5}$$

The proposed method generates 210 security properties (30 DSPs and 180 ISPs) w.r.t. the *design security rules*. It detects 16 security properties violation including two DSPs and 14 ISPs. The main reason for this security problem is the weak security policy of AMBA-2.0 AHB. The only policy implemented in AMBA-2.0 AHB is that for receiving transactions generated by master IPs, it checks whether or not the transactions' address is in the range of memory addresses. We fixed this security gap in AMBA-2.0 AHB by adding access control policies restricting the access of unauthorized master IPs to the secure memories. This can be done by checking the address of the received transactions whether or not they satisfy the expected range of addresses w.r.t. the design security rules. The properties were satisfied on next analysis run. A more general solution is to add a memory management unit to the AMBA-2.0 AHB.

For instance, the V_{AP} set related to the violated DSP is as follows:

$$V_{AP} = \{P14, P21 \mid$$

$$P14 = (0x69C450\_4 \rightarrow ahbin1 :: init1 :: gen\_frame \rightarrow READ$$

$$\rightarrow AHBMem2 :: trg2 :: exec\_func \rightarrow 0xA0005BA7 \rightarrow 4 \rightarrow 6071ns),$$

$$P21 = (0x6A9B20\_2 \rightarrow ahbin2 :: init2 :: gen\_frame \rightarrow READ$$

$$\rightarrow AHBMem2 :: trg2 :: exec\_func \rightarrow 0xA0005BB1 \rightarrow 4 \rightarrow 8043ns)$$

$$(6.6)$$

Path $P14$ in (6.6) shows that instance $init1$ of initiator module $ahbin1$ generates a transaction using gen_frame function to read from memory address 0xA0005BA7 referring to instance $trg2$ of target module $AHBMem2$. The target module AHB-$Mem2$ sends the response using $exec\_func$ at simulation time 6071 ns. In a same way, path $P21$ demonstrates that instance $init2$ of initiator module $ahbin2$ generates a transaction using gen_frame function to read from memory address 0xA0005BB1 referring to instance $trg2$ of target module $AHBMem2$. The target module $AHBMem2$ sends the response using $exec\_func$ at simulation time 8043 ns. The leakage depth analysis on V_{AP} in (6.6) shows that for both paths $P14$ and $P21$, the parameter $adrs$ is equal to 0xA0005BA7 and 0xA0005BB1, respectively. The parameter L is equal to 4 bytes for all paths.

A part of the V_{AP} set regarding the ISP is as follows:

$$V_{AP} = \{(P411, \ P532, \ P565) \mid$$

$$P411 = (0x1DBCD00\_21 \rightarrow AES\_core :: aes\_master :: gen\_frame$$

$$\rightarrow READ \rightarrow Mem\_Secure1 :: mem\_sec1 :: exec\_func$$

$$\rightarrow 0xB0000010 \rightarrow 4 \rightarrow 999550ns),$$

$$P532 = (0x1DBCD00\_54 \rightarrow AES\_core :: aes\_master :: gen\_frame$$

$$\rightarrow WRITE \rightarrow AHBMem2 :: trg2 :: exec\_func$$

$$\rightarrow 0xA0000BA9 \rightarrow 4 \rightarrow 1007190ns)$$

$$P565 = (0x6A9B20\_19 \rightarrow ahbin2 :: init2 :: gen\_frame$$

$$\rightarrow READ \rightarrow AHBMem2 :: trg2 :: exec\_func\_func$$

$$\rightarrow 0xA0000BA9 \rightarrow 4 \rightarrow 1008680ns)\}$$

$$(6.7)$$

As illustrated in (6.7), a combination of paths *P411*, *P532*, and *P565* creating an indirect data flow between *ahbin2* and *Mem_Secure1* that is against the expected security rules. In path *P411* instance, *aes_master* of initiator module *AES_core* creates a transaction to read from memory address 0xB0000010 referring to the secure target module *Mem_Secure1* using *gen_frame* function. In path *P532*, the *AES_core* generates a transaction to write in memory address 0xA0000BA9 referring to target module *AHBMem2* when the simulation time advanced from 999550 to 1007190 ns. Finally, in path *P565* instance, *init2* of initiator module *ahbin2* reads from memory address 0xA0000BA9 at simulation time 1008680 ns.

For this experiment, the number of extracted transactions is 1100, and the whole analysis takes about 28 min to report the results.

6.3   Summary

In this chapter, we proposed two novel approaches that leverage information flow analysis to enable early security validation of VPs.

First, we presented the first VP-based IFT approach for security validation. At the heart of the approach is a scalable static information flow analysis that operates directly on the SystemC VP models. The analysis performs in interleaved manner connectivity analysis, access control extraction, call-graph analysis, data flow analysis, and static taint tracking to identify static paths that violate specified secure information flow properties. These potential vulnerable paths are reported back to user for further inspection. We have demonstrated the applicability of the approach on real-world VP SOCRocket.

Second, we presented the first dynamic VP-based IFT method for security validation. The IFT is performed by programming a debugger to automatically and non-intrusively extract the run-time behavior (TLM transactions) of a given SOC. The extracted transactions and the design security rules are automatically translated into a set of access paths and security properties, respectively. The proposed method validates the generated access paths against the security properties and reports back the vulnerable paths (that violate specified confidential information flow properties) to user for further inspection. The method is able to detect the exact point and the amount of information that is leaked. Experimental results confirm the applicability of our approach on various VP models including the real-world VP SOCRocket.

Chapter 7
Conclusion

In the last decade, IOT has significantly altered the requirements for heterogeneous SOCs. Tight integration of analog and digital IPs on a single die, while running software on top, has significantly increased the functionality and reduced the area of the SOC. However, the increased design complexity has become a bottleneck for the successful co-design of secure multi-disciplinary heterogeneous SOCs exhibiting tight interactions between HW/SW systems and their analog physical environment. As TTM becomes shorter, the ability to model and simulate complex heterogeneous SOCs where digital HW/SW is functionally intertwined with AMS IPs becomes more and more essential. If such overall system and architectural level models are available as early as possible in the design cycle, the architecture exploration and design issues, as well as security leaks, can be dramatically reduced.

To solve this problem, VP-based design and verification flow are heavily used nowadays for heterogeneous SOCs. It leverages the *Shifting Left* concept to enable HW/SW co-design. As a consequence, an executable description is made available, which is used as a golden reference for both (early) embedded SW development and HW verification. Hence, the functional correctness and security validation of VPs is very important, and the whole VP and its individual components, i.e., high-speed RF, AMS, and digital IPs, are subjected to rigorous verification. However, this modern VP-based design flow still has shortcomings, in particular due to the significant manual effort involved for verification and analysis tasks that is both time consuming and error prone. This book proposed several novel approaches to strongly enhance the modern VP-based verification flow of heterogeneous SOCs. In particular, the book contributions have been divided into four areas that will be briefly summarized in the following:

1. AMS Metamorphic Testing Environment
2. AMS Enhanced Code Coverage Verification Environment
3. AMS Enhanced Functional Coverage Verification Environment
4. Digital Early Security Validation

The first contribution improves the modern verification flow for AMS VPs by introducing a novel methodology leveraging *Metamorphic Testing* for SystemC/AMS-based designs. First, this book proposed a novel MT-approach for the verification of *linear* and *non-linear* behaviors of RF amplifiers at system level. It identified a set of high quality MRs to detect non-trivial bugs without the need of reference models. However, RF amplifiers are analog devices, and clearly, this is not sufficient as the true complexity stems from interactions between digital and analog signals in AMS systems. Therefore, this book broadened the MT-approach to verify complex AMS systems, in particular an industrial PLL. With the help of a set of high quality MRs, PLL behavior was verified at component- and system-level encompassing analog-to-digital, digital-to-analog, and digital-to-digital behaviors. As an advantage, the MT-based approach complements regression testing as the reference models are not needed to ensure correctness for different versions of VP.

The second contribution enhances the modern verification flow for AMS VPs by proposing novel code coverage closure methodologies. It resulted in an increase of up to 30% in verification quality of the VP. In particular, the book proposed a methodology for SW test qualification of IP integration in a software driven verification flow. It leveraged *mutation analysis* and basic code coverage metrics to enhance the verification quality of the VP. Albeit improving quality of VP, it falls short when considering the interactions between different variables. Hence, alternative approaches are required, which complement SDV. Therefore, this book also considered stronger coverage metrics such as *Data Flow Testing* (DFT) techniques that rely on test-case generation and simulation. It proposed DFT approaches for SystemC/AMS-based VPs and SystemC/AMS specific coverage criteria. As a result, high quality testbench and thoroughly verified VPs are obtained.

The third contribution are the enhancements in functional coverage closure methodologies for AMS systems, in particular RF amplifiers. First, the book proposed an AMS functional coverage-driven verification approach that elevated the concepts of digital functional coverage to the SystemC AMS abstraction. The approach is very effective to systematically verify the RF amplifiers; however, it has two shortcomings: (1) manual analysis (2) and unable to detect unstable behaviors of RF amplifiers. Therefore, this book also proposed a *Lightweight Coverage-Directed Stimuli Generation* (LCDG) approach that goes beyond *linear/non-linear* behaviors of AMS systems, i.e., it expanded CDG to verify the unstable behaviors of RF amplifiers at system level. It used simple *Euclidean Distance* instead of complex probabilistic models to achieve coverage closure. This makes the proposed approach *lightweight*. The proposed enhancements achieved a significant improvement of up to 30% in verification quality.

The fourth and the final contribution of this book are two approaches to ensure security validation of VPs at system level. The first approach is a scalable static information flow analysis that operates directly on the SystemC VP models. The analysis performs in interleaved data flow analysis and static taint tracking to identify static paths that violate specified secure information flow properties. While this approach is very effective in finding vulnerabilities, sometimes security leaks occur at run-time. Hence, extra protection is required. Therefore, the book also proposed a dynamic information flow analysis approach. The approach is performed by programming a debugger to automatically and non-intrusively extract the run-time behavior (TLM transactions) of a given VP.

All approaches have been implemented, discussed, and extensively evaluated with several experiments.[1] In summary, these contributions significantly enhance the modern VP-based verification flow by drastically improving the verification quality and at the same time reduce the overall verification effort due to the extensive automation.

7.1 Future Directions

The obtained results clearly demonstrate the effectiveness and benefits of the proposed approaches. Nonetheless, there is still room for additional improvement and extension. In particular, the following four directions are very promising and important to further boost and complement this book work:

- Investigate and validate the MT-approach at SPICE-level using the same MRs identified at the system level. As a consequence, a cross-level verification of DUV can be carried out. One-time effort is required to create the MRs; hence, the total verification time will be reduced. Furthermore, the reference models will not be needed. Furthermore, the MT-approach introduced in Chap. 3 can be expanded to more design classes to verify their behaviors. Another interesting direction is to devise solutions for determining the completeness of a set of MRs.
- Leverage code coverage metrics to detect code coverage anomalies in AMS VPs. Achieving 100% code coverage is in general impractical due to coverage anomalies, e.g., dead code—code that is not executed by any testsuite; hence, techniques are required to identify such anomalies. One direction in this regard is the dead code analysis that statically identifies the parts of DUV that are not used at all. Furthermore, these code coverage metrics in combination with slicing techniques can be used to localize bugs.
- Explore different techniques to improve and extend functional coverage-driven verification toward AMS. In particular, the neighboring events' identification process in Sect. 5.4 can be improved. Some promising directions are: (1) semantic similarity to the coverage goal, (2) based on goal's structure (e.g., dissimilarity of at least one dimension in cross-product coverage model), (3) use of static analysis techniques to identify goals that must be hit before selected goal is hit, etc. Lastly, the process of resolution estimation can be revised by casting it into an optimization problem with the goal to find the weights of static parameters.

---

[1] For our most recent VP-based approaches, visit www.systemc-verification.org.

- Consider security aspects in an AMS VP-based design flow to enable early security validation. In particular, data integrity and confidentiality properties and side channels are an important area. Given the widespread use of analog relies on a complex business model, wherein the functionality (i.e., physical interfaces, sensors, actuators, wireless communications, and so on) in most contemporary systems, there is an alarming lack of understanding and an urgent need for a comprehensive study of the threat and solution space in the AMS/RF domain. Furthermore, covertly stealing sensitive information through hardware Trojans embedded in analog/RF ICs is another direction that can be extended.

Bibliography

1. Accellera, SystemC analog/mixed-signal user's guide (2020)
2. Accellera Systems Initiative (2016). http://www.accellera.org/downloads/standards/systemc
3. H. Agrawal, R.A. DeMillo, B. Hathaway, W. Hsu, W. Hsu, E.W. Krauser, R.J. Martin, A.P. Mathur, E. Spafford, Design of mutant operators for the C programming language. Technical report, Purdue University, 1989
4. R.T. Alexander, J. Offutt, A. Stefik, Testing coupling relationships in object-oriented programs. Softw. Testing Verif. Reliab. **20**(4), 291–327 (2010)
5. Analog Devices. Fundamentals of phase locked loops (PLLs). *MT–086 Tutorial* (2009)
6. Apple. *Small Chip. Giant Leap* (Apple Inc., 2020)
7. J. Aynsley, TLM-2.0 base protocol checker. https://www.doulos.com/knowhow/systemc/tlm2. Accessed 30 Jan 2018
8. E. Barke, D. Grabowski, H. Graeb, L. Hedrich, S. Heinen, R. Popp, S. Steinhorst, Y. Wang, Formal approaches to analog circuit verification, in *2009 Design, Automation & Test in Europe Conference & Exhibition* (IEEE, Piscataway, 2009), pp. 724–729
9. M. Barnasconi, SystemC AMS extensions: solving the need for speed (2010)
10. M. Barnasconi, K. Einwich, C. Grimm, T. Maehne, A. Vachoux, Advancing the SystemC analog/mixed-signal (AMS) extensions. *Open SystemC Initiative* (2011)
11. M. Barnasconi, K. Einwich, C. Grimm, T. Maehne, A. Vachoux, et al., Standard SystemC AMS extensions 2.0 language reference manual. *Accellera Systems Initiative* (2013)
12. M. Barnasconi, C. Grimm, M. Damm, K. Einwich, M. Louërat, T. Maehne, F. Pecheux, A. Vachoux, SystemC AMS extensions user's guide. *Accellera Systems Initiative* (2010)
13. M. Barnasconi, C. Grimm, M. Damm, K. Einwich, M. Louërat, T. Maehne, F. Pecheux, A. Vachoux, SystemC AMS extensions user's guide. *Accellera Systems Initiative* (2010)
14. J.S. Barros, V.H. Schulz, D.V. Lettnin, An adaptive closed-loop verification approach in UVM-SystemC for AMS circuits, in *2018 31st Symposium on Integrated Circuits and Systems Design* (IEEE, Piscataway, 2018), pp. 1–6
15. N. Bombieri, F. Fummi, V. Guarnieri, G. Pravadelli, F. Stefanni, T. Ghasempouri, M. Lora, G. Auditore, M.N. Marcigaglia, Reusing RTL assertion checkers for verification of SystemC TLM models. J. Electron. Testing Theory Appl. **31**(2), 167–180 (2015)
16. N. Bombieri, F. Fummi, G. Pravadelli, A mutation model for the SystemC TLM 2.0 communication interfaces, in *Design, Automation and Test in Europe* (2008), pp. 396–401
17. N. Bombieri, F. Fummi, G. Pravadelli, On the mutation analysis of SystemC TLM-2.0 standard, in *IEEE International Workshop on Microprocessor Test and Verification* (2009), pp. 32–37

© The Author(s), under exclusive license to Springer Nature Switzerland AG 2023
M. Hassan et al., *Enhanced Virtual Prototyping for Heterogeneous Systems*,
https://doi.org/10.1007/978-3-031-05574-4

18. T.A. Budd, R.J. Lipton, R. DeMillo, F. Sayward, The design of a prototype mutation system for program testing, in *AFIPS* (1978), pp. 623–627
19. J.E. Chen, Modeling RF systems. *The Designer's Guide Community*, 2005
20. T.Y. Chen, F.-C. Kuo, H. Liu, P.-L. Poon, D. Towey, T. Tse, Z.Q. Zhou, Metamorphic testing: a review of challenges and opportunities. ACM Comput. Surv. **51**(1), 4 (2018)
21. T.Y. Chen, F.-C. Kuo, Y. Liu, A. Tang, Metamorphic testing and testing with special values, in *International Conference on Software Engineering, Artificial Intelligence, Networking, and Parallel/Distributed Computing* (2004), pp. 128–134
22. C.-N. Chou, Y.-S. Ho, C. Hsieh, C.-Y. Huang, Symbolic model checking on SystemC designs, in *Design Automation Conf.* (2012), pp. 327–333
23. M. Cieplucha, Metric-driven verification methodology with regression management. J. Electron. Testing **35**(1), 101–110 (2019)
24. A. Cimatti, I. Narasamdya, M. Roveri, Software model checking SystemC. IEEE Trans. Comput. Aided Design Circuits Syst. **32**(5), 774–787 (2013)
25. COSEDA Technologies GmbH. PLL example for COSIDE® 2.5. https://www.coseda-tech. com/files/coside/user_files/Files/PLL_Example_for_COSIDE_2.5.zip
26. R.A. DeMillo, R.J. Lipton, F.G. Sayward, Hints on test data selection: help for the practicing programmer. IEEE Comput. **11**(4), 34–41 (1978)
27. G. Denaro, A. Margara, M. Pezzè, M. Vivanti, Dynamic data flow testing of object oriented systems, in *International Conference on Software Engineering* (2015), pp. 947–958
28. A. Devices, Relative humidity measurement system. http://www.analog.com/media/en/ reference-design-documentation/reference-designs/CN0346.pdf
29. A. Dias, Phase locked loop simulator in SystemC-AMS. https://americodias.com/docs/ systemc-ams/pll.md
30. H. Do, G. Rothermel, On the use of mutation faults in empirical assessments of test case prioritization techniques. IEEE Trans. Softw. Eng. **32**(9), 733–752 (2006)
31. A.F. Donaldson, A. Lascu, Metamorphic testing for (graphics) compilers, in *Proceedings of the 1st International Workshop on Metamorphic Testing* (2016), pp. 44–47
32. Z.J. Dong, M.H. Zaki, G. Al Sammane, S. Tahar, G. Bois, Checking properties of PLL designs using run-time verification, in *International Conference on Microelectronics* (2007), pp. 125–128
33. K. Einwich, Introduction to the SystemC AMS extension standard, in *IEEE Workshop on Design and Diagnostics of Electronic Circuits and Systems* (2011), pp. 6–8
34. A. Ferraiuolo, R. Xu, D. Zhang, A.C. Myers, G.E. Suh, Verification of a practical hardware security architecture through static information flow analysis. ACM SIGARCH Comput. Archit. News **45**, 555–568 (2017)
35. T. Ferrere, *Assertions and measurements for mixed-signal simulation*. Ph.D. Thesis, Université Grenoble-Alpes, France, 2016
36. S. Fine, A. Ziv, Coverage directed test generation for functional verification using Bayesian networks, in *Design Automation Conf.* (2003)
37. A. Fürtig, G. Gläser, C. Grimm, L. Hedrich, S. Heinen, H.-S. L. Lee, G. Nitsche, M. Olbrich, C. Radojicic, F. Speicher, Novel metrics for Analog Mixed-Signal coverage, in *IEEE Workshop on Design and Diagnostics of Electronic Circuits and Systems* (2017), pp. 97–102
38. F.M. Gardner, *Phaselock Techniques* (Wiley, London, 2005)
39. M. Goli, R. Drechsler, *Automated Analysis of Virtual Prototypes at the Electronic System Level: Design Understanding and Applications* (Springer Nature, 2020)
40. M. Goli, M. Hassan, D. Große, R. Drechsler, Security validation of VP-based SoCs using dynamic information flow tracking. Inform. Technol. **61**(1), 45–58 (2019)
41. M. Goli, J. Stoppe, R. Drechsler, Automatic protocol compliance checking of SystemC TLM-2.0 simulation behavior using timed automata, in *Int'l Conf. on Comp. Design* (2017)
42. M. Goli, J. Stoppe, R. Drechsler, Automated nonintrusive analysis of electronic system level designs. IEEE Trans. Comput.-Aided Design Integr. Circuits Syst. **39**(2), 492–505 (2020)

43. C. Grimm, M. Barnasconi, A. Vachoux, K. Einwich, An introduction to modeling embedded analog/mixed-signal systems using SystemC AMS extensions, in *Design Automation Conf.*, vol. 23 (2008).

44. C. Grimm, M. Barnasconi, A. Vachoux, K. Einwich, An introduction to modeling embedded analog/mixed-signal systems using SystemC AMS extensions, in *Open SystemC Initiative* (2008)

45. D. Große, R. Drechsler, *Quality-Driven SystemC Design* (Springer, Berlin, 2010)

46. D. Große, U. Kühne, R. Drechsler, Analyzing functional coverage in bounded model checking. IEEE Trans. Comput. Aided Design Circuits Syst. **27**(7), 1305–1314 (2008)

47. D. Große, H.M. Le, M. Hassan, R. Drechsler, Guided lightweight software test qualification for IP integration using virtual prototypes, in *Int'l Conf. on Comp. Design* (2016), pp. 606–613

48. B.J. Grun, D. Schuler, A. Zeller, The impact of equivalent mutants, in *International Conference on Software Testing, Verification, and Validation Workshops* (2009), pp. 192–199

49. S. Gupta, B.H. Krogh, R.A. Rutenbar, Towards formal verification of analog designs, in *International Conference on Computer-Aided Design* (2004), pp. 210–217

50. F. Haedicke, H.M. Le, D. Große, R. Drechsler, CRAVE: An advanced constrained random verification environment for SystemC, in *International Symposium on System-on-Chip* (2012), pp. 1–7

51. D. Haerle, Trends and challenges in Analog and Mixed-Signal verification, in *Frontiers in Analog CAD Keynote Address* (2018)

52. R.G. Hamlet, Testing programs with the aid of a compiler. IEEE Trans. Softw. Eng. **SE-3**(4), 279–290 (1977)

53. M. Hampton, S. Petithomme, Leveraging a commercial mutation analysis tool for research, in *Testing: Academic and Industrial Conference Practice and Research Techniques— MUTATION* (2007), pp. 203–209

54. W. Hartong, N. Luetke-Steinhorst, R. Schweiger, Coverage driven verification for mixed signal systems, in *ANALOG Conference* (2008)

55. M. Hassan, D. Große, R. Drechsler, System-level verification of linear and non-linear behaviors of RF amplifiers using metamorphic relations, in *ASP Design Automation Conf.* (2021)

56. M. Hassan, D. Große, R. Drechsler, System level verification of phase-locked loop using metamorphic relations, in *Design, Automation and Test in Europe* (2021)

57. M. Hassan, D. Große, H.M. Le, R. Drechsler, Data flow testing for SystemC-AMS timed data flow models, in *Design, Automation and Test in Europe* (2019), pp. 366–371

58. M. Hassan, D. Große, H.M. Le, T. Vörtler, K. Einwich, R. Drechsler, Testbench qualification for SystemC-AMS timed data flow models, in *Design, Automation and Test in Europe* (2018), pp. 857–860

59. M. Hassan, D. Große, T. Vörtler, K. Einwich, R. Drechsler, Functional coverage-driven characterization of RF amplifiers, in *Forum on Specification and Design Languages* (2019), pp. 1–8

60. M. Hassan, V. Herdt, H.M. Le, M. Chen, D. Große, R. Drechsler, Data flow testing for virtual prototypes, in *Design, Automation and Test in Europe* (2017), pp. 380–385

61. M. Hassan, V. Herdt, H.M. Le, D. Große, R. Drechsler, Early SoC security validation by VP-based static information flow analysis, in *International Conference on Computer-Aided Design* (2017), pp. 400–407

62. J. Havlicek, S. Little, Realtime regular expressions for analog and mixed-signal assertions, in *Int'l Conf. on Formal Methods in CAD* (2011), pp. 155–162

63. V. Herdt, D. Große, R. Drechsler, Closing the RISC-V compliance gap: looking from the negative testing side, in *Design Automation Conf.* (2020)

64. V. Herdt, D. Große, R. Drechsler, *Enhanced Virtual Prototyping: Featuring RISC-V Case Studies* (Springer, Berlin, 2020)

65. V. Herdt, D. Große, R. Drechsler, Fast and accurate performance evaluation for RISC-V using virtual prototypes, in *Design, Automation and Test in Europe* (2020), pp. 618–621

66. V. Herdt, D. Große, R. Drechsler, RVX—a tool for concolic testing of embedded binaries targeting RISC-V platforms, in *International Symposium on Automated Technology for Verification and Analysis* (Springer, Berlin, 2020), pp. 543–549
67. V. Herdt, D. Große, R. Drechsler, Towards specification and testing of RISC-V ISA compliance, in *Design, Automation and Test in Europe* (2020), pp. 995–998
68. V. Herdt, D. Große, R. Drechsler, C. Gerum, A. Jung, J.-J. Benz, O. Bringmann, M. Schwarz, D. Stoffel, W. Kunz, Systematic RISC-V based firmware design, in *Forum on Specification and Design Languages* (2019), pp. 1–8
69. V. Herdt, H.M. Le, R. Drechsler, Verifying SystemC using stateful symbolic simulation, in *Design Automation Conf.* (2015), pp. 49:1–49:6
70. V. Herdt, H.M. Le, D. Große, R. Drechsler, Compiled symbolic simulation for SystemC, in *International Conference on Computer-Aided Design* (2016), pp. 52:1–52:8
71. IEEE Std. 1666, *IEEE Standard SystemC LRM* (2011)
72. IEEE Std. 1800, *IEEE SystemVerilog* (2005)
73. O.S. Initiative et al., IEEE standard SystemC language reference manual. IEEE Comput. Soc. 1666–2005 (2006)
74. D.C. Jarman, Z.Q. Zhou, T.Y. Chen, Metamorphic testing for Adobe data analytics software, in *International Workshop on Metamorphic Testing* (IEEE, Piscataway, 2017), pp. 21–27
75. Y. Jia, M. Harman, An analysis and survey of the development of mutation testing. IEEE Trans. Softw. Eng. **37**(5), 649–678 (2011)
76. H. Kai, P. Zhu, R. Yan, X. Yan, Functional testbench qualification by mutation analysis. VLSI Design **2015**, 256474:1–256474:9 (2015)
77. K. Karnane, G. Curtis, R. Goering, Solutions for mixed-signal SoC verification, in *Cadence Design Systems* (2009)
78. A.G. Kimura, K.-W. Liu, S. Prabhu, S.B. Bibyk, G. Creech, Trusted verification test bench development for phase-locked loop (PLL) hardware insertion, in *IEEE International Midwest Symposium on Circuits and Systems* (IEEE, Piscataway, 2013), pp. 1208–1211
79. P. Kocher, J. Horn, A. Fogh, D. Genkin, D. Gruss, W. Haas, M. Hamburg, M. Lipp, S. Mangard, T. Prescher, et al., Spectre attacks: exploiting speculative execution, in *2019 IEEE Symposium on Security and Privacy (SP)* (2019), pp. 1–19
80. D. Kulkarni, A.N. Fisher, C.J. Myers, A new assertion property language for analog/mixed-signal circuits, in *Forum on Specification and Design Languages* (2013)
81. K. Kundert, Accurate and rapid measurement of IP2 and IP3. *The Designer's Guide Community*, 2002
82. J.W. Laski, B. Korel, A data flow oriented program testing strategy. IEEE Trans. Softw. Eng. **9**(3), 347–354 (1983)
83. C. Lattner, LLVM and Clang: next generation compiler technology, in *The BSD Conference* (2008), pp. 1–2
84. H.M. Le, D. Große, R. Drechsler, Automatic TLM fault localization for SystemC. IEEE Trans. Comput. Aided Design Circuits Syst. **31**(8), 1249–1262 (2012)
85. H.M. Le, D. Große, V. Herdt, R. Drechsler, Verifying SystemC using an intermediate verification language and symbolic simulation, in *Design Automation Conf.* (2013), pp. 116:1–116:6
86. H.M. Le, V. Herdt, D. Große, R. Drechsler, Towards formal verification of real-world SystemC TLM peripheral models—a case study, in *Design, Automation and Test in Europe* (2016), pp. 1160–1163
87. E. Lefeuvre, D. Audigier, C. Richard, D. Guyomar, Buck-boost converter for sensorless power optimization of piezoelectric energy harvester. IEEE Trans. Power Electron. **22**(5), 2018–2025 (2007)
88. R. Leupers, F. Schirrmeister, G. Martin, T. Kogel, R. Plyaskin, A. Herkersdorf, M. Vaupel, Virtual platforms: breaking new grounds, in *Design, Automation and Test in Europe* (2012), pp. 685–690

89. M. Lipp, M. Schwarz, D. Gruss, T. Prescher, W. Haas, A. Fogh, J. Horn, S. Mangard, P. Kocher, D. Genkin, et al., Meltdown: reading kernel memory from user space 973–990 (2018)
90. L. Liu, S. Vasudevan, Efficient validation input generation in RTL by hybridized source code analysis, in *Design, Automation and Test in Europe* (2011), pp. 1596–1601
91. M. Lora, S. Vinco, E. Fraccaroli, D. Quaglia, F. Fummi, Analog models manipulation for effective integration in smart system virtual platforms. IEEE Trans. Comput. Aided Design Circuits Syst. **37**(2), 378–391 (2018)
92. O. Maler, D. Ničković, Monitoring properties of analog and mixed-signal circuits. Softw. Tools Technol. Transf. **15**(3), 247–268 (2013)
93. Mentor Graphics. Coverage Cookbook, 2013. https://verificationacademy.com/cookbook/coverage
94. F. Pêcheux, C. Grimm, T. Maehne, M. Barnasconi, K. Einwich, SystemC AMS based frameworks for virtual prototyping of heterogeneous systems, in *IEEE International Symposium on Circuits and Systems* (2018), pp. 1–4
95. R. Pelánek, Fighting state space explosion: Review and evaluation, in *International Workshop on Formal Methods for Industrial Critical Systems* (Springer, Berlin, 2008), pp. 37–52
96. A. Piziali, *Functional Verification Coverage Measurement and Analysis* (Springer, Berlin, 2007)
97. C. Radojicic, F. Schupfer, M. Rathmair, C. Grimm, Assertion-based verification of signal processing systems with affine arithmetic, in *Forum on Specification and Design Languages* (2012), pp. 20–26
98. S. Rapps, E.J. Weyuker, Selecting software test data using data flow information. IEEE Trans. Softw. Eng. **11**(4), 367–375 (1985)
99. B.C. Schafer, A. Mahapatra, S2CBench: Synthesizable SystemC benchmark suite for high-level. IEEE Embedded Syst. Lett. **6**(3), 53–56 (2014)
100. T. Schuster, R. Meyer, R. Buchty, L. Fossati, M. Berekovic, SoCRocket—a virtual platform for the European Space Agency's SoC development, in *International Symposium on Reconfigurable and Communication-Centric Systems-on-Chip* (2014), pp. 1–7. http://github.com/socrocket
101. S. Segura, G. Fraser, A.B. Sanchez, A. Ruiz-Cortés, A survey on metamorphic testing. IEEE Trans. Softw. Eng. **42**(9), 805–824 (2016)
102. S. Segura, Z.Q. Zhou, Metamorphic testing 20 years later: a hands-on introduction, in *International Conference on Software Engineering* (2018), pp. 538–539
103. A. Sen, Mutation operators for concurrent SystemC designs, in *IEEE International Workshop on Microprocessor Test and Verification* (2009), pp. 27–31
104. A. Sen, Concurrency-oriented verification and coverage of system-level designs. ACM Trans. Design Autom. Electron. Syst. **16**(4), 37 (2011)
105. Y. Serrestou, V. Beroulle, C. Robach, Functional verification of RTL designs driven by mutation testing metrics, in *EUROMICRO Symposium on Digital System Design* (2007), pp. 222–227
106. S. Simon, D. Bhat, A. Rath, J. Kirscher, L. Maurer, Coverage-driven mixed-signal verification of smart power ICS in a UVM environment, in *European Test Symposium* (IEEE, Piscataway, 2017), pp. 1–6
107. S. Simon, G. Pelz, L. Maurer, Accelerating coverage collection for mixed-signal systems in a UVM environment, in *Forum on Specification and Design Languages* (2015), p. 31
108. S. Skorobogatov, C. Woods, Breakthrough silicon scanning discovers backdoor in military chip, in *International Workshop on Cryptographic Hardware and Embedded Systems* (Springer, Berlin, 2012), pp. 23–40
109. S. Steinhorst, L. Hedrich, Formal methods for verification of analog circuits, in *Simulation and Verification of Electronic and Biological Systems* (Springer, Berlin, 2011), pp. 173–192
110. S. Steinhorst, L. Hedrich, Equivalence checking of nonlinear analog circuits for hierarchical AMS system verification, in *2012 IEEE/IFIP 20th International Conference on VLSI and System-on-Chip (VLSI-SoC)* (IEEE, Piscataway, 2012), pp. 135–140

111. T. Su, Z. Fu, G. Pu, J. He, Z. Su, Combining symbolic execution and model checking for data flow testing, in *International Conference on Software Engineering* (2015), pp. 654–665
112. Synopsys. Certitude, 2015. https://www.synopsys.com/Tools/Verification/FunctionalVerification/Pages/certitude-ds.aspx
113. Q. Tao, W. Wu, C. Zhao, W. Shen, An automatic testing approach for compiler based on metamorphic testing technique, in *ASP Design Automation Conf.* (2010), pp. 270–279
114. S. Tasiran, K. Keutzer, Coverage metrics for functional validation of hardware designs. IEEE Design Test Comput. **18**(4), 36–45 (2001)
115. C. Technologies. COSIDE®. http://www.coseda-tech.com
116. Texas Instruments. Disentangle RF amplifier specs: output voltage/current and 1db compression point, 2016
117. M. Vivanti, A. Mis, A. Gorla, G. Fraser, Search-based data-flow test generation, in *International Symposium on Software Reliability Engineering* (2013), pp. 370–379
118. P. Vizmuller, *RF Design Guide: Systems, Circuits, and Equations* (Artech House, 1995)
119. T. Vörtler, K. Einwich, M. Hassan, D. Große, Using constraints for SystemC AMS design and verification, in *Design and Verification Conference & Exhibition Europe* (2018)
120. D. Walter, S. Little, C. Myers, N. Seegmiller, T. Yoneda, Verification of analog/mixed-signal circuits using symbolic methods. IEEE Trans. Comput. Aided Design Circuits Syst. **27**(12), 2223–2235 (2008)
121. M.H. Zaki, S. Tahar, G. Bois, Formal verification of analog and mixed signal designs: a survey. Microelectron. J. **39**(12), 1395–1404 (2008)
122. C. Zivkovic, C. Grimm, Symbolic simulation of SystemC AMS without yet another compiler, in *Forum on Specification and Design Languages* (2018), pp. 5–16
123. C. Zivkovic, C. Grimm, Nubolic simulation of AMS systems with data flow and discrete event models, in *Design, Automation and Test in Europe* (2019)

Index

Printed in the United States
by Baker & Taylor Publisher Services